Springer Proceedings in Complexity

Springer Complexity

Springer Complexity is an interdisciplinary program publishing the best research and academic-level teaching on both fundamental and applied aspects of complex systems—cutting across all traditional disciplines of the natural and life sciences, engineering, economics, medicine, neuroscience, social, and computer science.

Complex Systems are systems that comprise many interacting parts with the ability to generate a new quality of macroscopic collective behavior the manifestations of which are the spontaneous formation of distinctive temporal, spatial, or functional structures. Models of such systems can be successfully mapped onto quite diverse "real-life" situations like the climate, the coherent emission of light from lasers, chemical reaction–diffusion systems, biological cellular networks, the dynamics of stock markets and of the Internet, earthquake statistics and prediction, freeway traffic, the human brain, or the formation of opinions in social systems, to name just some of the popular applications.

Although their scope and methodologies overlap somewhat, one can distinguish the following main concepts and tools: self-organization, nonlinear dynamics, synergetics, turbulence, dynamical systems, catastrophes, instabilities, stochastic processes, chaos, graphs and networks, cellular automata, adaptive systems, genetic algorithms, and computational intelligence.

The three major book publication platforms of the Springer Complexity program are the monograph series "Understanding Complex Systems" focusing on the various applications of complexity, the "Springer Series in Synergetics", which is devoted to the quantitative theoretical and methodological foundations, and the "SpringerBriefs in Complexity" which are concise and topical working reports, case-studies, surveys, essays, and lecture notes of relevance to the field. In addition to the books in these two core series, the program also incorporates individual titles ranging from textbooks to major reference works.

More information about this series at http://www.springer.com/series/11637

Bruno Gonçalves · Ronaldo Menezes
Roberta Sinatra · Vinko Zlatic
Editors

Complex Networks VIII

Proceedings of the 8th Conference on Complex Networks CompleNet 2017

Springer

Editors
Bruno Gonçalves
Center for Data Science
New York University
New York, NY
USA

Roberta Sinatra
Center for Network Science and
Mathematics Department
Central European University
Budapest
Hungary

Ronaldo Menezes
BioComplex Laboratory, School
of Computing
Florida Institute of Technology
Melbourne, FL
USA

Vinko Zlatic
Rudjer Bošković Institute
Theoretical Physics Division
Zagreb
Croatia

ISSN 2213-8684 ISSN 2213-8692 (electronic)
Springer Proceedings in Complexity
ISBN 978-3-319-54240-9 ISBN 978-3-319-54241-6 (eBook)
DOI 10.1007/978-3-319-54241-6

Library of Congress Control Number: 2017932410

Printed on acid-free paper

This Springer imprint is published by Springer Nature
The registered company is Springer International Publishing AG
The registered company address is: Gewerbestrasse 11, 6330 Cham, Switzerland

Preface

The International Workshop on Complex Networks CompleNet (www.complenet.org) was initially proposed in 2008, and the first workshop took place in 2009 in Catania. The initiative was the result of efforts from researchers from the (i) BioComplex Laboratory in the Department of Computer Sciences at Florida Institute of Technology, USA, and the (ii) Dipartimento di Ingegneria Informatica e delle Telecomunicazioni, University di Catania, Italy. CompleNet aims at bringing together researchers and practitioners working on complex networks or related areas. In the past two decades, we have indeed witnessed an exponential increase of the number of publications in this field. From biology to computer science, from economics to social systems, complex networks are becoming pervasive in many fields of science. CompleNet aims at addressing this interdisciplinary nature of complex networks. CompleNet 2017 was the eighth event in the series and was hosted at the Inter University Center Dubrovnik, Croatia, during March 21–24, 2017.

This book includes the peer-reviewed list of works presented at CompleNet 2017. We received 106 submissions from 32 countries. Each submission was reviewed by at least three members of the Program Committee. Acceptance was judged based on the relevance to the symposium themes, clarity of presentation, originality and accuracy of results and proposed solutions. After the review process, 9 full papers and 13 short papers were selected to be included in this book. The 22 contributions in this book address many topics related to complex networks and have been organized in seven major groups: (1) Theory of complex networks, (2) Community detection, (3) Dynamics and spreading phenomena on networks, (4) Applications of network science, (5) Social structure, (6) Human behavior, (7) Biological networks. We would like to thank the Program Committee members for their work in promoting the event and refereeing submissions. We are grateful to our speakers: Johan Bollen, Guido Caldarelli, Gourab Ghoshal, Aniko Hannak, Ágnes Horvát, Vito Latora, Jörg Menche, Stasa Milojevic, Anastasios Noulas, Giovanni Petri, Zoltan Toroczkai; their presentation is one of the reasons CompleNet 2017 was such a success.

New York, NY, USA — Bruno Gonçalves
Melbourne, FL, USA — Ronaldo Menezes
Budapest, Hungary — Roberta Sinatra
Zagreb, Croatia — Vinko Zlatic

Contents

Part VI Human Behavior

Contributors

Talel Abdessalem Computer Science and Networks Department, Telecom ParisTech, Paris, France

Mohammed Abufouda Computer Science Department, University of Kaiserslautern, Kaiserslautern, Germany

Eric F.M. Araújo VU University Amsterdam, Amsterdam, The Netherlands

A.L.C. Bazzan Instituto de Informática da UFRGS, Porto Alegre, Brazil

Olga Berestneva National Research Tomsk Polytechnic University, Tomsk, Russia

Alessandro Bessi Information Sciences Institute, University of Southern California, Los Angeles, CA, USA

Livio Bioglio Department of Computer Science, University of Turin, Turin, Italy

Pierre Borgnat CNRS, Laboratoire de Physique, Univ Lyon, Ens de Lyon, Univ Claude Bernard, Villeurbanne, France

Remy Cazabet Sorbonne Universites, UPMC Univ Paris 06, Paris, France

Nazim Choudhury Faculty of Engineering and IT, Centre for Complex Systems Research, The University of Sydney, Redfern, NSW, Australia

Ingemar J. Cox Department of Computer Science, University College London (UCL), London, UK

Jean Creusefond GREYC, Normandie Université, Caen, France

S.R. Dahmen Instituto de Física da UFRGS, Porto Alegre, Brazil

Alexandre P. Francisco INESC-ID/Instituto Superior Técnico, Universidade de Lisboa, Lisbon, Portugal

Ralucca Gera Department of Applied Mathematics, Naval Postgraduate School, Monterey, CA, USA

Bruno Gonçalves Center for Data Science, New York University, New York, NY, USA

R. Gramsch Historisches Institut der Universität Jena, Jena, Germany

Emre Guney Joint IRB-BSC-CRG Program in Computational Biology, Institute for Research in Biomedicine (IRB), Barcelona, Spain

Aart T. van Halteren VU University Amsterdam, Amsterdam, The Netherlands; Philips Research, Eindhoven, The Netherlands

Harith Hamoodat BioComplex Laboratory, School of Computing Florida Institute of Technology, Melbourne, USA

Lars K. Hansen Department of Applied Mathematics and Computer Science, Danish Technical University (DTU), Kongens Lyngby, Denmark

D. Ienco IRSTEA - UMR TETIS, Montpellier, France

Margarita Ifti Faculty of Natural Sciences, University of Tirana, Tirana, Albania

R. Interdonato DIMES - University of Calabria, Rende, Italy

Pablo Jensen CNRS, Laboratoire de Physique, Univ Lyon, Ens de Lyon, Univ Claude Bernard, Villeurbanne, France

Nikolay Kinash Irkutsk National Research Technical University, Irkutsk, Russia

Michel C.A. Klein VU University Amsterdam, Amsterdam, The Netherlands

Inva Koçiaj Faculty of Natural Sciences, University of Tirana, Tirana, Albania

Adnan Manzoor VU University Amsterdam, Amsterdam, The Netherlands

Malvina Marku Faculty of Natural Sciences, University of Tirana, Tirana, Albania

Ronaldo Menezes BioComplex Laboratory, School of Computing, Florida Institute of Technology, Melbourne, USA

Shigeyuki Miyagi The University of Shiga Prefecture, Hikone, Shiga, Japan

Yasutaka Mizui The University of Shiga Prefecture, Hikone, Shiga, Japan

Julia S. Mollee VU University Amsterdam, Amsterdam, The Netherlands

Terrence J. Moore U.S. Army Research Lab, Adelphi, USA

Tsuyoshi Murata Department of Computer Science, School of Computing, Tokyo Institute of Technology, Tokyo, Japan

Klotilda Nikaj Faculty of Natural Sciences, University of Tirana, Tirana, Albania

Kyosuke Nobuto The University of Shiga Prefecture, Hikone, Shiga, Japan

Vatsal Ojha Dougherty Valley High School, San Ramon, CA, USA

Siddharth Pal Raytheon BBN Technologies, Cambridge, USA

Pedro Paredes CRACS & INESC-TEC, DCC-FCUP, Universidade do Porto, Porto, Portugal

Ruggero G. Pensa Department of Computer Science, University of Turin, Turin, Italy

Fabio Petroni Sapienza University of Rome, Rome, Italy

P. Poncelet LIRMM - Université de Montpellier, Montpellier, France

Walter Quattrociocchi IMT Institute for Advanced Studies, Lucca, Italy

Ram Ramanathan Raytheon BBN Technologies, Cambridge, USA

Pedro Ribeiro CRACS & INESC-TEC, DCC-FCUP, Universidade do Porto, Porto, Portugal

Alessandra Rossodivita Luigi Sacco Academic Hospital, Milan, Italy

Younis Al Rozz BioComplex Laboratory, School of Computing Florida Institute of Technology, Melbourne, USA

Osamu Sakai The University of Shiga Prefecture, Hikone, Shiga, Japan

A. Sallaberry LIRMM - Université Paul Valéry, Montpellier, France

Francisco C. Santos INESC-ID/Instituto Superior Técnico, Universidade de Lisboa, Lisbon, Portugal

Antonio Scala ISC CNR, Rome, Italy

S.M. Mahdi Seyednezhad BioComplex Laboratory, School of Computing, Florida Institute of Technology, Melbourne, USA

Miguel E.P. Silva CRACS & INESC-TEC, DCC-FCUP, Universidade do Porto, Porto, Portugal

Sucheta Soundarajan Department of Electrical Engineering & Computer Science, Syracuse University, New York, USA

Takahiko Sugihara Department of Computer Science, School of Computing, Tokyo Institute of Technology, Tokyo, Japan

Ananthram Swami U.S. Army Research Lab, Adelphi, USA

A. Tagarelli DIMES - University of Calabria, Rende, Italy

Sude Tavassoli Graph Theory and Complex Network Analysis Group, Computer Science Department, Kaiserslautern University of Technology, Kaiserslautern, Germany

Andreia Sofia Teixeira INESC-ID/Instituto Superior Técnico, Universidade de Lisboa, Lisbon, Portugal

Alexei Tikhomirov Inha University, Incheon, Republic of Korea

Andrey Trufanov Irkutsk National Research Technical University, Irkutsk, Russia

Shahadat Uddin Faculty of Engineering and IT, Centre for Complex Systems Research, The University of Sydney, Redfern, NSW, Australia

Michela Del Vicario IMT Institute for Advanced Studies, Lucca, Italy

Pivithuru Wijegunawardana Department of Electrical Engineering & Computer Science, Syracuse University, New York, USA

Shi Zhou Department of Computer Science, University College London (UCL), London, UK

Fabiana Zollo IMT Institute for Advanced Studies, Lucca, Italy

Katharina A. Zweig Graph Theory and Complex Network Analysis Group, Computer Science Department, Kaiserslautern University of Technology, Kaiserslautern, Germany

Part I
Theory of Complex Networks

Second-Order Assortative Mixing in Social Networks

Shi Zhou, Ingemar J. Cox and Lars K. Hansen

Abstract In a social network, the number of links of a node, or node degree, is often assumed as a proxy for the node's importance or prominence within the network. It is known that social networks exhibit the (first-order) assortative mixing, i.e. if two nodes are connected, they tend to have similar node degrees, suggesting that people tend to mix with those of comparable prominence. In this paper, we report the *second-order* assortative mixing in social networks. If two nodes are connected, we measure the degree correlation between their most prominent *neighbours*, rather than between the two nodes themselves. We observe very strong second-order assortative mixing in social networks, often significantly stronger than the first-order assortative mixing. This suggests that if two people interact in a social network, then the importance of the most prominent person each knows is very likely to be the same. This is also true if we measure the average prominence of neighbours of the two people. This property is weaker or negative in non-social networks. We investigate a number of possible explanations for this property. However, none of them was found to provide an adequate explanation. We therefore conclude that second-order assortative mixing is a new property of social networks.

1 Background

A network or graph consists of nodes connected together via links. Networks are utilised in many disciplines. The nodes model physical elements such as people, proteins or cities, and the links between nodes represent connections between them, such as contacts, biochemical interactions, and roads. In recent years studying the

S. Zhou (✉) · I.J. Cox
Department of Computer Science, University College London (UCL),
Gower Street, London WC1E 6BT, UK
e-mail: s.zhou@ucl.ac.uk

L.K. Hansen
Department of Applied Mathematics and Computer Science, Danish Technical
University (DTU), Kongens Lyngby, Denmark

B. Gonçalves et al. (eds.), *Complex Networks VIII*,
Springer Proceedings in Complexity, DOI 10.1007/978-3-319-54241-6_1

structure, function and evolution of networked systems in society and nature has become a major research focus [2, 4, 7, 19, 21, 23].

The degree, k, of a node is defined as the number of links the node possesses. The probability distribution of node degrees is indicative of a network's global connectivity. For example random graphs with a Poisson degree distribution [9] have most nodes with degrees close to the average degree. In contrast, many complex networks in nature and society are scale-free graphs [1] exhibiting a power-law degree distribution, where many nodes have only a few links and a small number of nodes have very large numbers of links. However, the degree distribution alone does not provide a full description of a network's topology. Networks with exactly the same degree distribution can possess other properties that are vastly different [13, 15, 24].

One such property, is the mixing pattern between the two end nodes of a link [17, 18], i.e. the joint probability distribution of a node with degree k being connected to a node with degree k'. In general, biological and technological networks are *disassortative* mixing meaning that well-connected nodes tend to link with poorly-connected nodes, and vice versa. In contrast, social networks, such as collaborations between film actors or scientists, exhibit *assortative* mixing, where nodes with similar degrees tend to be connected.

To quantify this mixing property, Newman [17] proposed the assortative coefficient, r, where $-1 \leq r \leq 1$. It is derived by considering the Pearson correlation between two sequences, where corresponding elements in the two sequences represent the degree of the nodes at either end of a link in the network. For a directed network, the degree of the starting node of a link is contained in one sequence, and the degree of the ending node is in the other sequence. The number of elements in each sequence is the number of links. For an undirected network, as all the networks studied in this paper, each undirected link is replaced by two directed links pointing at opposite directions. Thus the number of elements in a sequence is twice the number of links.

A network with assortative mixing is characterised by a possible value of r; where $r = 1$ corresponds to a perfect assortative mixing, i.e., every link connects two nodes with the same degree. A network with disassortative mixing has a negative value of r; where $r = -1$ corresponds to a perfect disassortative mixing, i.e., every link connects two nodes with difference degrees. When r equals or close to 0, there is no degree correlation, i.e., the network is random or neutral in terms of degree mixing.

The mixing pattern has been studied as a fundamental property of networks, and the assortative coefficient r has been widely used to measure this property.

2 Second-Order Mixing Pattern

We now introduce and define a related property which we refer to as the *second-order* mixing pattern.

2.1 Definition of $\mathscr{R}_{max}$ and $\mathscr{R}_{avg}$

Following Newman's definition of the (first-order) assortative coefficient r [17], we define $\mathscr{R}_{max}$ as the second-order assortative coefficient based on the neighbours maximum degree,

$$\mathscr{R}_{max} = \frac{L^{-1}\sum_i K_i K_i' - \left[\frac{1}{2}L^{-1}\sum_i (K_i + K_i')\right]^2}{\frac{1}{2}L^{-1}\sum_i (K_i^2 + K_i'^2) - \left[\frac{1}{2}L^{-1}\sum_i (K_i + K_i')\right]^2}, \tag{1}$$

where L is the number of links in the network, K_i and K_i' are the neighbours maximum degrees of the two nodes a and b connected by the link i, i.e. $K_i = \max(k_{n_a} : n_a \in N_{a\backslash b})$ and $K_i' = \max(k_{n_b} : n_b \in N_{b\backslash a})$; $N_{a\backslash b}$ denotes the set of neighbours of node a, excluding node b; and $N_{b\backslash a}$ denotes the set of neighbours of node b, excluding node a.

Note that when calculating the first and second assortative coefficients, we actually use the *excess* degree [17], which is degree minus one. This is because, when considering two connected nodes, A and B, the neighbourhood of A is defined to exclude B. And likewise for the neighbourhood of B. See Fig. 1 for examples.

Similarly we define $\mathscr{R}_{avg}$ as the second-order assortative coefficient based on the neighbours average degrees by replacing K_i and K_i' in the above equation as $K_i = \frac{1}{k_a}\sum_{n_a \in N_{a\backslash b}} k_{n_a}$ and $K_i' = \frac{1}{k_b}\sum_{n_b \in N_{b\backslash a}} k_{n_b}$.

2.2 Results

We consider eleven networks, including five social networks, two biological networks, two technology networks, and two synthetic networks based on random connections [9] and the Barabási and Albert [1] model, respectively. Values of both

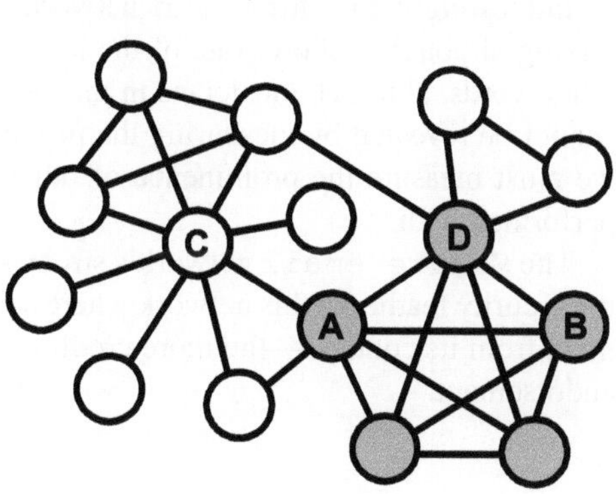

Fig. 1 Examples of node excess degrees, which is node degree minus one. Consider the link between nodes A and B, the excess degree of node A is 5; and the neighbours maximum excess degree of node A is 7, which is the excess degree of node C

the first-order and the second-order assortative coefficients, r, $\mathscr{R}_{avg}$ and $\mathscr{R}_{max}$ are provided in Table 1.

2.2.1 Statistical Significance

The expected standard deviation σ on the value of assortative coefficient r can be obtained by the jackknife method [8] as $\sigma^2 = \sum_{i=1}^{L}(r_i - r)^2$, where r_i is the value of r for the network in which the i-th link is removed and $i = 1, 2, ...L$. And likewise for second-order assortative coefficients $\mathscr{R}_{max}$ and $\mathscr{R}_{avg}$. For all cases shown in Table 1, the value of σ is very small (<0.03), which validates the statistical significance of the coefficients.

2.2.2 Null Hypothesis Test

A high correlation score between two value sequences must be tested against the null hypothesis. For each network and each coefficient in Table 1, we randomly permuted the order of degree values in one of the two degree sequences and re-computed the coefficient. This was repeated 100 times and then we calculated the mean and standard deviation. Our calculation shows that for each network and each coefficient the mean value is close to zero and the standard deviation is small. This result again confirms the statistical significance of the first and second-order assortative coefficients.

2.2.3 Social Networks

Four social networks (a)-(d) show positive values of first-order assortative coefficient, and notably they show significantly higher values of second-order assortative coefficients. This indicates that in these social networks, people judge on other individual's social status based on not only the individual's own prominence (e.g. the number of co-starred films or co-authored publications), but more crucially, the prominence of is collaborators.

Interestingly, the `Musician` network exhibits very low first-order assortative mixing although it shows one of the strongest second-order assortative mixing. In other words, although musicians in this network do exhibit a strong social parity, it cannot be revealed by measuring the prominence of musicians themselves; instead we must measure the prominence of other musicians that each musician has ever performed with.

The `Secure email` network's strong second-order assortative mixing is due to the security feather of this network where a person's security credit relies on endorsement from its contacts—the more credit a contact already has the more valuable its endorsement.

Table 1 Properties of the networks under study. Properties shown are the numbers of nodes and links; the assortative coefficients r, $\mathscr{R}_{avg}$ and $\mathscr{R}_{max}$ with the corresponding expected standard deviation σ_r, σ_{avg} and σ_{max}; and the average clustering coefficient of nodes in a network, $\langle C \rangle$. (**a**) Film actor collaborations [1], where two actors are connected if they have co-starred in a film; (**b**) Scientist collaborations [16], where two scientists are connected if they have co-authored a paper in condense matter physics; (**c**) Jazz musician network [10], where two musicians are connected if they have played in a band; (**d**) Secure email network [3], where a link represent a secure email exchange between two trusted users using the Pretty Good Privacy (PGP) algorithm; (**e**) General email network [11], where email exchanges take place at a university, including a large amount of unsolicited emails; (**f**) Western States Power Grid of the United States [22]; (**g**) *C. elegans* metabolic network [12], where two metabolites are connected if they participate in a biochemical reaction; (**h**) the protein interactions of the yeast *Saccharomyces cerevisiae* [5, 14]; (**i**) Internet [18] (http://www.routeviews.org/), where two service providers are connected if they have a commercial agreement to exchange data traffic; (**j**) the random graphs [9]; and (**k**) the Barabási-Albert (BA) graphs [1]

Network	Nodes	Links	r	σ_r	$\mathscr{R}_{avg}$	σ_{avg}	$\mathscr{R}_{max}$	σ_{max}	$\langle C \rangle$
(a) Film actor	82,593	3,666,738	0.206	0.013	0.836	0.027	0.813	0.009	0.75
(b) Scientist	12,722	39,967	0.161	0.007	0.680	0.014	0.647	0.005	0.65
(c) Musician	198	2,742	0.020	0.019	0.543	0.023	0.307	0.029	0.62
(d) Secure email	10,680	24,316	0.238	0.007	0.653	0.009	0.680	0.007	0.27
(e) General email	1,133	5,451	0.078	0.014	0.242	0.014	0.247	0.014	0.22
(f) Power grid	4,941	6,594	0.004	0.014	0.205	0.015	0.258	0.016	0.08
(g) Metabolism	453	2,025	−0.226	0.011	0.265	0.032	0.263	0.023	0.65
(h) Protein	4,626	14,801	−0.137	0.008	−0.046	0.007	0.033	0.009	0.09
(i) Internet	11,174	23,409	−0.195	0.001	−0.097	0.004	0.036	0.008	0.30
(j) Random graph	10,000	30,000	$\simeq 0$	0.009	$\simeq 0$	0.011	$\simeq 0$	0.006	$\simeq 0$
(k) BA graph	10,000	30,000	$\simeq 0$	0.004	$\simeq 0$	0.008	$\simeq 0$	0.008	$\simeq 0$

2.2.4 Non-social Networks

Other forms of networks do not exhibit the very strong second-order correlations exhibited by social networks.

The `General email` network is not considered as a typical social network, because it contains a large amount of unsolicited, one-way communications, such as notices and advertisements forwarded from departmental secretaries to all students. Not surprisingly, this network's second assortative mixing is as weak as the `Metabolism` and `Power grid` networks.

The `Internet` and `Protein` networks, the second order assortative coefficients are either zero (R_{max}) or negative (R_{avg}).

As expected, neutral random networks generated by graph models are completely uncorrelated, i.e. $R_{max} = R_{avg} = 0$.

2.3 *Frequency Distributions of Links as Functions of Degrees*

Figure 2 provides a more detailed look into the assortative mixing in the `Scientist` network. Figure 2a shows there is a strong first-order correlation between node degrees when $k < 20$, and the correlation rapidly decreases with increasing degree, as expected for a scale-free network.

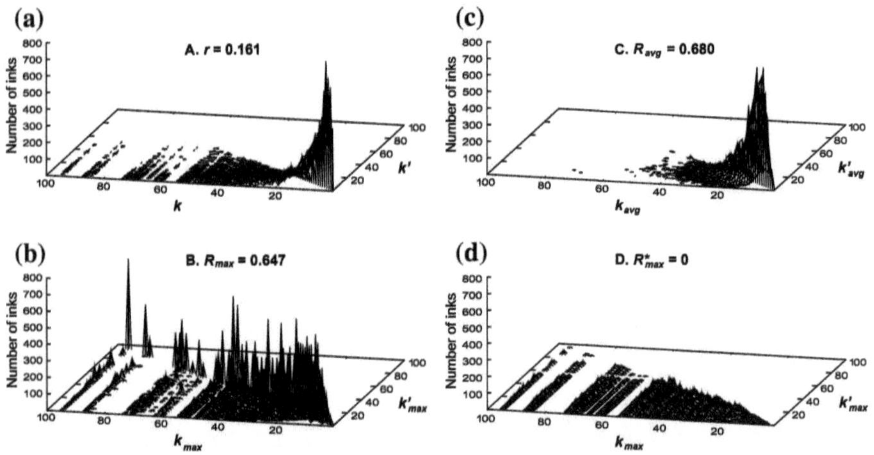

Fig. 2 The first and second-order assortative mixing in the `Scientist` network. We show the link frequency distribution as functions of (*A*) degrees k and k' of the two end nodes of a link, with $k \geq k'$; (*B*) the neighbours maximum degrees of the two nodes, K_{max} and K'_{max}, with $K_{max} \geq K'_{max}$; and (*C*) the neighbours average degrees, K_{avg} and K'_{avg}, with $K_{avg} \geq K'_{avg}$, respectively. (*D*) is the same as (B), where links are randomly rewired while preserving the degree distribution. The maximum degree of the network is 97

For the second-order assortative mixing, Fig. 2b shows a very strong correlation for almost all values of the neighbours maximum degree K_{max}, where the link distribution along the diagonal does not decrease with the increase of K_{max}. Of course the correlation in Fig. 2b is not perfect, and a second process appears to be uniform noise. The noise might be better modelled as Gaussian which is probably due to the summation of many nodes and the central limit theorem. If the neighbours average degree rather than the maximum is considered, we still observe a strong correlation in Fig. 2c.

3 Seeking Possible Explanations

Here we examine whether the second-order mixing is a *new* topological property, i.e. whether it can be explained by other known properties of the networks.

3.1 Increased Neighbourhood

One may wonder whether the strong correlation scores associated with second-order assortative mixing could simply be due to the increased neighbourhood (from distance of one hop to two hops), as a node always has more second-order neighbours than first-order neighbours. To exclude this possibility we also examined the Xth-order assortative coefficients, $\mathscr{R}_{max}$ and $\mathscr{R}_{avg}$, which are calculated using the maximum or average degree within the neighbourhood of up to X hops from each end node of a link. Of course, if the neighbourhood continues to increase, we observed that eventually the coefficients would increase and approach to one. This is to be expected since eventually, the neighbourhood encompasses the entire network.

However, we observed that for *all* networks under study, the values of the third-order coefficients were actually smaller than the 2nd-order coefficients. This suggest that the second-order assortative mixing cannot be explained by increased neighbourhood.

The fact that the third-order coefficients are smaller than the second-order coefficients has rich meanings. For technology networks, consider the Internet, where a network service provider only cares about the prominence of a customer (disassortative first-order mixing), it does not know and care about who else the customer has linked with (neutral second-order mixing), and care even less about those one step further away. For social networks, one tends to match its collaborator's prominence (first-order assortative mixing) and the prominence of the collaborator's contacts (stronger second-order assortative mixing), but it does not know or care about contacts of the collaborator's contacts whom the collaborator does not know directly. In other words, the value of social prominence vanishes rapidly after the second-order.

3.2 High-Degree Nodes

Another possible explanation for the high values of second-order assortative coefficients considered, is that there are a few hub nodes that are extremely well connected and dominate the network structure. To test this we removed the best-connected node (together with the links attaching to it and any resulting isolated nodes) from the networks and re-computed the coefficients. We also calculate the coefficients after removing the top 5 best-connected nodes. Results show that in all cases, the coefficients change very little. For some networks, such as the `Secure email`, `Musician` and `Metabolism` networks, the second-order coefficients became stronger after the best-connected nodes are removed. This suggest that the second-order assortative mixing cannot be explained by the existent of high-degree nodes.

3.3 Power-Law Degree Distribution

While high degree nodes do not explain the high second order assortative mixing scores, the underlying heterogenous power-law structure of the networks was also a possible explanation. To exclude this possibility we used the random link rewiring algorithm [15, 24] to produce surrogate networks by randomly rewiring links while preserving the exact degree distribution of the networks under study.

Figure 2d illustrates the distribution of links as a function of K_{max} and K'_{max} in a randomly rewired version of the `Scientist` network. The second-order assortative mixing in the original network disappears completely in the randomised case.

This result shows that the second-order mixing is determined by a network's degree distribution, because two networks (the original and the randomised case) with the identical degree distribution show hugely different mixing patterns, both in the first-order [15, 24] and in the second-order (see Fig. 2).

This result again demonstrates the limitation of characterising network topology by degree distribution alone, and highlights the critical importance of characterising a network's topology using multiple properties from different aspects.

3.4 Clustering Coefficient

We also examined whether the second-order assortative mixing is a consequence of the clustering behaviour observed in many social networks, where one's friends are also friends of each other. This is quantified by the clustering coefficient, C_i, which is defined as $C_i = \frac{e_i}{k_i(k_i-1)/2}$, where k_i is the degree of node i and e_i is the number of connections between the node's neighbours [23]. The average clustering coefficient, $\langle C \rangle$, is the arithmetic average over all nodes in the network. Comparison of $\langle C \rangle$ against $\mathscr{R}_{avg}$ and $\mathscr{R}_{max}$ in Table 1 and Fig. 3 shows that high values of the

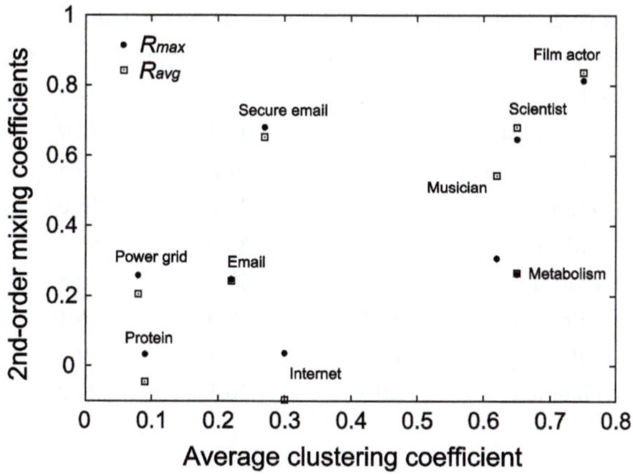

Fig. 3 Second-order mixing coefficients vs average clustering coefficient

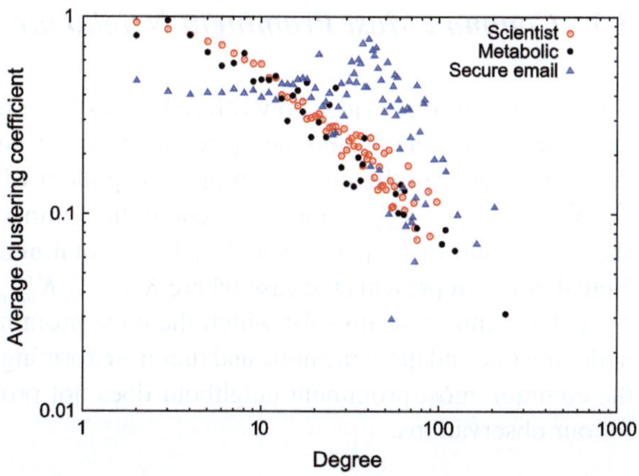

Fig. 4 Average clustering coefficient of *k*-degree nodes

second-order coefficients occur for both high and low values of clustering coefficient. There is no correlation between them.

Figure 4 reveals that the `Scientist` network and the `Secure email` network are fundamentally different in the relation between clustering coefficient and node degree, yet they have similar $\mathscr{R}_{avg}$ and $\mathscr{R}_{max}$. Whereas the `Scientist` network and the `Metabolism` exhibit very similar clustering coefficient properties, but their second-order coefficients are significantly different.

The above results suggest that the second-order assortative mixing is something quite unexpected, particularly considering the work on the hierarchical organisation of complex networks [6, 20].

Table 2 Link ratio values of the networks under study

Network	$L_{<4}/L$ (%)	$L_{<2}/L$ (%)	$L_{<1}/L$ (%)	L_{Δ}/L (%)	R_{max}
(a) Film actor	34.2	34.2	34.1	34.1	0.813
(b) Scientist	50.2	45.4	42.9	41.6	0.647
(c) Secure email	43.2	37.8	35.5	34.6	0.680
(d) Email	26.8	20.2	15.0	12.8	0.247
(e) Musician	56.3	56.0	55.7	55.7	0.307
(f) Metabolism	51.3	51.1	50.8	50.7	0.263
(g) Protein	10.5	8.0	6.4	5.7	0.033
(h) Power grid	68.1	38.3	19.1	7.8	0.258
(i) Internet	20.5	20.3	20.2	20.2	0.036

3.5 Common Most Prominent Neighbour

It is interesting to consider how often the most prominent contact at each end of a link is the same person, and therefore they form a triangle. Let X denote the degree difference between the most prominent neighbour of the two end nodes of a link, i.e. $X = |K_{max} - K'_{max}|$, and $L_{<x}$ denote the number of links with $X < x$. Table 2 shows the ratio of $L_{<4}$, $L_{<2}$ and $L_{<1}$ to the total number of links, L, respectively. Note that $L_{<1}$ represents the case where $K_{max} = K'_{max}$. Also shown is L_{Δ}/L, where L_{Δ} is the number of links for which the most prominent neighbour of the two end nodes are one and the same node and therefore forming a triangle, $L_{\Delta} \in L_{<1}$. Clearly the common most prominent neighbour does not provide an adequate explanation for our observations.

3.6 Bipartite Network

A bipartite network is a network with two non-overlapping sets of nodes Δ and Γ, where all links must have one end node belonging to each set. For example, actors star in films, scientist write papers, and musician play in bands. The `Film actor`, `Scientist` and `Musician` networks under study are constructed from bipartite networks, e.g. two actors are linked if they co-star in a film and two scientists are linked if they co-author a paper.

The `Film actor`, `Scientist` and `Musician` networks all exhibit strong second-order assortative mixing. It is therefore reasonable to ask whether the second-order assortative mixing can be attributed to the nature of bipartite networks? For example, all actors of one film constitute a complete subgraph, in which everyone connects with the highest-degree node in the group.

However, we found no support for this hypothesis. Firstly, the `Metabolic` network is also constructed from a bipartite network where the two types of nodes are

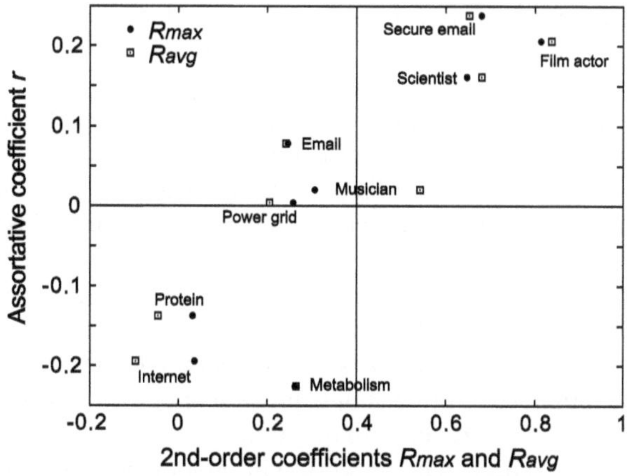

Fig. 5 Second-order assortative coefficients R_{max} and R_{avg} vs first-order assortative coefficient r

metabolites and reactions. Two metabolites are linked if they participate in a reaction. The `Metabolic` network, however, does not show a strong second-order assortative mixing.

Secondly, the `Secure email` network is a non-bipartite network, where two email users are linked by direct email communications. It exhibits one of the strongest second-order assortative mixing.

3.7 Relation Between First and Second-Order Mixing Coefficients

Figure 5 compares the assortative coefficient r and the second-order coefficients R_{max} and R_{avg} for the networks under study. They are seemingly loosely related.

However, there are exceptions. Consider the `Metabolic` network and the `Email` network, the former is strongly disassortative with $r = -0.226$, whereas the later is assortative with $r = 0.078$. Yet both networks exhibit similar values of the second-order mixing coefficients.

4 Conclusion

Our experimental results demonstrated very strong-second order assortative mixing in social networks where human are in charge of forming connections; but weaker, or even negative values for biological and technological networks where there is a lack of social preference.

We examined a larger variety of other network properties in an effort to establish whether second-order assortative mixing was induced from other network properties such as its power law distribution, cluster coefficient, and bipartite graphs. However, although some of them might be a contributing factor, none of these properties was found to provide an adequate explanation. We therefore conclude that second-order assortative mixing is a new property, which reveals a new dimension to the hierarchical structure present in social networks.

For social networks, the degree of a node is often considered a proxy for the prominence or importance of a person. First order assortative mixing has then been interpreted as indicating that if two people interact in a social network then they are likely to have similar prominence. The much stronger second-order assortative mixing suggests that there could be an even stronger social parity when measuring the prominence of a person's contacts. Whether our most prominent contacts serve to introduce us or we simply prefer to mix with people who know similarly important people, remains an open question.

We expect that our work will provide new clues for studying the structure and evolution of social networks as well as complex networks in general.

Acknowledgements The authors thank Ole Winther and Sune Lehmann of DTU, Denmark for discussions relating to the clustering coefficient.

References

1. Barabási, A., Albert, R.: Emergence of scaling in random networks. Science **286**, 509 (1999)
2. Barabási, A.L.: Linked: The New Science of Networks. Perseus Publishing (2002)
3. Boguñá, M., Pastor-Satorras, R., Vespignani, A.: Cut-offs and finite size effects in scale-free networks. Eur. Phys. J. B **38**, 205–210 (2004)
4. Bornholdt, S., Schuster, H.G.: Handbook of Graphs and Networks—From the Genome to the Internet. Wiley-VCH, Weinheim Germany (2002)
5. Colizza, V., Flammini, A., Maritan, A., Vespignani, A.: Characterization and modeling of protein-protein interaction networks. Physica A **352**, 1–27 (2005)
6. Dorogovtsev, S.N., Goltsev, A.V., Mendes, J.F.F.: Pseudofractal scale-free web. Phys. Rev. E **65**, 066122 (2002)
7. Dorogovtsev, S.N., Mendes, J.F.F.: Evolution of Networks—From Biological Nets to the Internet and WWW. Oxford University Press, Oxford (2003)
8. Efron, B.: Computers and the theory of statistics thinking the unthinkable. SIAM Rev. **21**, 460–480 (1979)
9. Erdös, P., Rényi, A.: On random graphs. Publ. Math. Debrecen **6**, 290 (1959)
10. Gleiser, P., Danon, L.: Community structure in jazz. Adv. Complex Syst. **6**, 565–573 (2003)
11. Guimera, R., Danon, L., Diaz-Guilera, A., Giralt, F., Arenas, A.: Self-similar community structure in a network of human interactions. Phys. Rev. E **68**, 065103 (2003)
12. Jeong, H., Tombor, B., Albert, R., Oltvai, Z.N., Barabási, A.-L.: The large-scale organization of metabolic networks. Nature **407**, 651–654 (2000)
13. Mahadevan, P., Krioukov, D., Fall, K., Vahdat, A.: Systematic topology analysis and generation using degree correlations. ACM SIGCOMM Comput. Commun. Rev. **36**(4), 135–146 (2006)
14. Maslov, S., Sneppen, K.: Specificity and stability in topology of protein networks. Science **296**(5569), 910–913 (2002)
15. Maslov, S., Sneppenb, K., Zaliznyaka, A.: Detection of topological patterns in complex networks: correlation profile of the internet. Physica A **333**, 529–540 (2004)

16. Newman, M.: Scientific collaboration networks. I. Network construction and fundamental results. Phys. Rev. E **64**(016131), 016131 (2001)
17. Newman, M.E.J.: Assortative mixing in networks. Phys. Rev. Lett. **89**(208701), 208701 (2002)
18. Pastor-Satorras, R., Vázquez, A., Vespignani, A.: Dynamical and correlation properties of the Internet. Phys. Rev. Lett. **87**(258701), 258701 (2001)
19. Pastor-Satorras, R., Vespignani, A.: Evolution and Structure of the Internet—A Statistical Physics Approach. Cambridge University Press, Cambridge (2004)
20. Ravasz, E., Barabási, A.-L.: Hierarchical organization in complex networks. Phys. Rev. E **67**, 026112 (2003)
21. Wasserman, S., Faust, K.: Social Network Analysis. Cambridge University Press, Cambridge (1994)
22. Watts, D.J., Strogatz, S.H.: Collective dynamics of 'small-world' networks. Nature **393**, 440 (1998)
23. Watts, J.: Small Worlds: The Dynamics of Networks between Order and Randomness. Princeton Univeristy Press, New Jersey, USA (1999)
24. Zhou, S., Mondragón, R.: Structural constraints in complex networks. New J. Phys. **9**(173), 1–11 (2007)

Network Motifs Detection Using Random Networks with Prescribed Subgraph Frequencies

Miguel E.P. Silva, Pedro Paredes and Pedro Ribeiro

Abstract In order to detect network motifs we need to evaluate the exceptionality of subgraphs in a given network. This is usually done by comparing subgraph frequencies on both the original and an ensemble of random networks keeping certain structural properties. The classical null model implies preserving the degree sequence. In this paper our focus is on a richer model that approximately fixes the frequency of subgraphs of size $K-1$ to compute motifs of size K. We propose a method for generating random graphs under this model, and we provide algorithms for its efficient computation. We show empirical results of our proposed methodology on neurobiological networks, showcasing its efficiency and its differences when comparing to the traditional null model.

Keywords Network motifs · Random graphs · Subgraph counting

1 Introduction

Complex networks have been established as essential tools to model and analyze several real-life systems and problems. A technique that greatly contributed for this reputation is network motif analysis [15]. Network motifs consist of over-represented substructures of a network, or subgraphs that appear in a higher number than expected. This method has been used successfully in many fields of science, such as biology [22, 23] or sociology [4].

In order to perform a meaningful network motif analysis, it is important to decide on a definition of what is the expected frequency of a certain subgraph. To do so, one chooses a determined null model of random graphs and computes the average

M.E.P. Silva (✉) · P. Paredes · P. Ribeiro
CRACS & INESC-TEC, DCC-FCUP, Universidade do Porto, Porto, Portugal
e-mail: mepsilva@dcc.fc.up.pt

P. Paredes
e-mail: pparedes@dcc.fc.up.pt

P. Ribeiro
e-mail: pribeiro@dcc.fc.up.pt

B. Gonçalves et al. (eds.), *Complex Networks VIII*,
Springer Proceedings in Complexity, DOI 10.1007/978-3-319-54241-6_2

frequency of the given subgraph on this null model. The most used null model is maintaining the degree sequence of the original network [4, 13, 14, 23]. Other models have been proposed [3, 15], but here we focus on a new model.

One can think of graph edges as subgraphs of size 2. A natural extension would therefore be to maintain counts of larger subgraphs. Moreover, certain patterns can be essentially the consequence of over-represented smaller subgraphs contained in them. With all of this mind we propose to keep the frequency of subgraphs of size $K-1$ when discovering motifs of size K, aiming towards a much richer null model, able to really distinguish when a subgraph is really significant by itself and not just a product of smaller subtopologies. A limited version of this idea for size 4 motifs was shown in [15], but here we aim for a generic method (that works for any feasible K) and that is also efficient.

Our main contributions to the stated problem are the following:

- A method that generates random networks using the invariant of subgraphs of frequency $K-1$, up to a certain margin, with an algorithm based on simulated annealing [10];
- A study of different ways of applying the previous method by using additional invariants like the classic degree sequence invariant;
- An algorithm, based on [17, 25], that updates the frequency of subgraphs after an edge addition or removal, which is used in order to compute the frequencies of subgraphs of size $K-1$ that the mentioned method requires;

We analyze our method to show that it is both efficient and accurate. To do so, we rely on different real complex networks and show that our method obtains different results when comparing with the classic degree sequence model. We also show that our frequency update algorithm performs much better than recalculating all frequencies in every iteration of the generation method.

The rest of this paper is organized as follows. Section 2 discusses some preliminaries and background concepts regarding network motif analysis, needed for the following sections. Section 3 presents our generation method and also show some of its properties. In Sect. 4 we showcase our frequency updating algorithm and prove its correctness. Section 5 contains a brief experimental analysis of our proposed methods and algorithms. Finally we conclude in Sect. 6.

1.1 Related Work

Milo et al. [15] use, as null model, random graphs that maintain the degree sequence and subgraph count of size $K-1$, when calculating motifs of size K. Their implementation uses a Monte Carlo Metropolis-Hastings algorithm for directed networks to calculate motifs of size 4, but does not suggest an immediate strategy for undirected networks or subgraph size greater than 4.

In other related work, Bois and Gayraud in [3] use prior probability to generate random graphs with a given count of subgraphs, but only present priors for two types of directed subgraphs of size 3. Ritchie et al. [21] present an algorithm parametrized by a degree sequence and a set of subgraphs that generates random graphs with those parameters. It is based on the matching algorithm [14], whereas our work uses a Markov chain Monte Carlo method of generation.

We also note that, as far as we know, there is no known method that efficiently updates subgraph frequencies on an edge addition or removal.

2 Network Motif Finding

2.1 Definition of Network Motif

The concept of motifs as building blocks of networks was first described by Milo et al. in [15] as patterns of inter-connections occurring in numbers that are significantly higher than what one would expect. To simplify notation, we will refer to network motifs simply as motifs.

A determined subgraph is considered significant if its frequency in the original graph is exceptionally high in comparison with its frequency on random networks under a certain null model. To assess exceptionality, one computes the probability that the number of times the subgraph appears on a randomized network is lower than on the original network and then compares it with a certain threshold P. This probability can be estimated using Z-scores on a standard normal distribution, by computing the standardized difference between the observed and expected frequency.

To be classified as a motif, according to the original definition [15], it is also required to fulfill two other properties. For a given subgraph, let f_o be the frequency of the subgraph on the original network and f_r the average frequency of the same subgraph on random networks with an unspecified null model. The first constraint is minimal frequency, that is, f_o has to have a minimum value of U, to ensure a quantitative minimum. The second constraint is minimal deviation, that is, f_o needs to be significantly larger than f_r, to prevent the detection of motifs that have a small difference between these two values but have a narrow distribution in the random networks. This can be stated has $f_o - f_r > D \cdot f_r$, where D is a proportionality threshold.

With this information, we can give a formal definition of motif. Given a set of parameters $\{P, U, D\}$, a subgraph of a given graph is considered a motif if:

- $P(f_r > f_o) \leq P$ **(over-representation)**
- $f_o \geq U$ **(minimal frequency)**
- $f_o - f_r > Df_r$ **(minimum deviation)**

2.2 Algorithms for Subgraph Counting

The main primitive of motif finding is counting subgraphs on graphs, which is called a subgraph census. There are essentially three different ways of doing so: in a network centric way, which corresponds to counting the occurrences of all subgraphs up to a certain size K; in a subgraph centric way, which corresponds to counting the occurrences of a single subgraph; in a set centric way, which corresponds to counting the occurrences of a set of subgraphs.

The state of the art algorithms that do a generic network centric census are QuateXelero [9] and FaSE [17], which are similar contemporaneous algorithms. Both build on previous methods [25] that do an enumeration of all subgraphs up to a certain size K and then perform isomorphism tests on each one using a tool like `nauty` [11]. By building an intermediate structure (a quaternary tree and a g-trie, respectively) the number of necessary isomorphism tests is decreased to a multiple of the number of different types of subgraph present in the network. More recently, some methods [12, 18] explore combinatorial properties of graphs to achieve algorithms that are orders of magnitude better than any generic method, but that can only work with subgraphs up to a certain size (currently up to 5 for undirected graphs [18] and 4 for directed [12]).

The most well known subgraph centric algorithm is the work by Grochow and Kellis [8], which efficiently counts the frequency of a single subgraph using a set of generated symmetry breaking conditions. Finally, there is only one known set centric algorithm, the work by Ribeiro and Silva [20].

2.3 Random Graphs

The study of random graphs is growing rapidly as a model of complex networks. Although the research on this topic dates back to the late 1950s, where, in a series of publications, Paul Erdos and Alfréd Rényi [5, 6] introduce a model, known as Erdos-Rényi (ER). In this model, each pair of vertices is connected with an independent probability p. More recently, other models have been proposed that follow closely characteristics from real world networks. Among these, Watts and Strogatz [24], propose a model to generate smalls-world graphs, networks whose average path length grows proportionally to the logarithm of the number of nodes in the network, and Barabasi-Albert [1] introduce another model for scale-free graphs [2], where the degree distribution follows a power law.

When focusing on more local properties, random graphs using a given degree sequence have become one of the most studied models, after their widespread use as null model for network motifs discovery [13, 15]. There is a multitude of algorithms to generate this type of graphs, of which we highlight the main two:

- The switching method [19] uses a Markov chain, starting with an initial network with the desired degree sequence and carries out a series of Monte Carlo switches that preserve that sequence.
- The matching algorithm [16] is based on "stubs". Each vertex is assigned a set of edge extremities, either incoming or outgoing. For each of these stubs, the vertex tries to connect with another one with the opposite type of stub.

On their original work, Milo et al. [15] use as null model both the degree sequence and subgraph frequency of size 3. To achieve this, they use the switching method to preserve the degree sequence and a Monte Carlo Metropolis-Hastings algorithm to approximate the subgraph count of the referred size. The frequency vectors are updated using analytical expressions using the neighbours of the vertices used for the edge switch.

3 Generation of Random Graphs

In this section, we discuss a generator of random graphs, with the novelty of allowing the random networks to be generated with approximately the same frequency of subgraphs of size $K - 1$ as an original network. We also permit the graphs to maintain or vary their degree sequence. The generation procedure is split in two phases: *randomization* and *convergence*.

3.1 Randomization

We offer three ways of creating an initial network. The first two employ a Markov chain edge swapping technique like in [15] and the third is a classical ER model, with number of edges equal to the number of edges in the original network.

The two Markov chain algorithms we utilize are similar, they both start with a real network and perform edge switches. The first version, which maintains degree sequence, given different nodes A, B, C and D, with connections $A \rightarrow B$ and $C \rightarrow D$, removes these existing connections and adds the new edges, $A \rightarrow D$ and $C \rightarrow B$. Nodes are selected in a way that ensures the prior inexistence of these two new connections. We do not distinguish between single and double edges, considering double edges simply as two independent single ones. The undirected case is easily generalizable.

The second type of Markov chain edge swap modifies the out-degree sequence of the network, for directed networks, and both in and out-degree sequences, in undirected networks. Given different nodes A, B and C, we delete the connection $A \rightarrow B$ and annex the edge $C \rightarrow B$, reducing the out-degree of node A by 1, while incrementing C's by the same amount. As before, nodes are selected with the requirement that A is connected to B but C is not.

The difference between the initial graphs produced by these two Markov chain variants lies in the time taken to converge to the desired subgraph count, the first version requires a lesser number of iterations. However, both produce graphs with a similar level of *energy*. Given two vectors (V_1 and V_2) with the number of appearances of each type of subgraph, where Γ denotes the set of these subgraphs, in two different networks, we define *energy* as the distance between these two vectors and calculate it as:

$$e = \frac{\sum_{i \in \Gamma} \frac{|V_{1,i} - V_{2,i}|}{V_{1,i} + V_{2,i}}}{|\Gamma|}$$

We refer to the energy of a random network as the distance between its vector of subgraph frequency and the corresponding vector from the original network.

For both Markov chain schemes, we repeat the edge swapping process $\mathcal{O}(E)$ times, where E represents the number of edges in the graph. The constant used is diverse in the existing literature, so we studied how the energy varies in function of the number of switches applied to the original network. We observed that a higher number of switches does not lead to higher energy. It should be noted that energy is not the sole measure of how well a graph is randomized and a low number of switches may not cause enough impact on other measures.

3.2 Convergence

After generating the initial network, we start the process of switching edges to obtain a subgraph count close to that of the real network. The convergence phase stops when the energy reaches a certain tunable threshold, where energy equal to 0 means that the subgraph frequencies of the random network and the original network are the same. In this phase, we use simulated annealing [10].

Simulated annealing is a metaheuristic technique used to approximate the global optimum of a large search space. On a general case, on each iteration, the heuristic chooses a random neighbouring state of the current state and decides probabilistically between changing to the new state or staying in the current one. This process is repeated until a global optimum solution is found or a solution that differs from the optimum less than a given threshold.

In our implementation of the method, the neighbouring state is chosen using the edge swapping mechanism described previously. If our initial network was obtained through the ER model or the out-degree changing Markov chain method, the swap also uses the out-degree changing switch. Otherwise, if the degree sequence was maintained throughout the randomization process, we only perform the type of switch that preserves it.

In order to decide if the the new candidate graph is accepted, we use an acceptance probability function $P(e, e', t)$, where e represents the current graph's energy, e' the candidate graph's energy and t is a parameter that decays over time, called the *temperature*. We use the same acceptance function as in the original formulation by Kirkpatrick et al. in [10], if $e' < e$, we always accept the transition, otherwise, we accept it with probability $exp(\frac{e-e'}{t})$.

A feature of simulated annealing is the decreasing temperature over time. This forces the state to converge to an optimum as, with lower temperature, the probability of accepting a state with higher energy is lessened. Upon reaching a point in the computation where the temperature reaches 0, only states with lesser energy are accepted and the computation eventually stops. The rate at which the temperature decreases is called the cooling factor of the algorithm.

4 Updating Frequencies of Subgraphs

The main bottleneck of the method described in the previous section is computing the frequencies of subgraphs in every iteration, to estimate the energy of the current solution. In [15], an analogous operation was done recounting the frequencies of subgraphs after each iteration of their algorithm until convergence. Our approach avoids recomputing all of the frequencies by only considering the subgraphs that are changed by the addition or removal of a certain edge.

The base of our method is the `FaSE` [17] algorithm, which we will extend in order to only count subgraphs that touch a given edge. Firstly, we will briefly describe the algorithm.

4.1 FaSE Algorithm

The original `FaSE` algorithm enumerates all connected subgraphs of a given size K and in the end computes the isomorphism of some of the subgraphs. To avoid having to compute the isomorphism of all subgraphs, the algorithm partitions subgraphs into intermediate classes during the enumeration process. By requiring that all subgraphs in one of the intermediate classes are isomorphic, in the end we only need to compute one isomorphism test per class. This is done by encapsulating the topological features of the enumerating graph in a tree like data structure. Thus, we can divide the algorithm into two interleaved concepts: the enumeration and a tree data structure.

Enumeration: The enumeration step can be done using any algorithm that grows a set of connected vertices. The algorithm from [25], `ESU`, was chosen since it is simple, efficient and fulfills all the requirements. We will describe its functioning since it will be useful for the end of this section.

`ESU` works by enumerating all size K subgraphs exactly once. It does so by keeping two ordered sets of vertices: V_s, which represents the partial subgraph that is currently being enumerated; V_{ext}, which represents the set of vertices that can be added to V_s as a valid extension. Each vertex is represented by a label which is unique and defined between 1 and $|V|$.

For each vertex v the algorithm repeats the same procedure setting initially $V_s = \{v\}$ and $V_{ext} = N(v)$, where $N(v)$ are the neighbors of v. This procedure starts by removing one element u of V_{ext} at a time. For each u, a new V'_s and V'_{ext} are created and the same procedure is repeated. V'_s is set to $V_s \cup \{u\}$ and V'_{ext} is set to V_{ext} without u and with additionally each element in $N_{exc}(u, V_s)$ with value greater than v. $N_{exc}(u, V_s)$ are the exclusive neighbors of u given V_s, that is, the neighbors of u that are not neighbors of elements in V_s. This procedure stops when the size of V_s reaches K, in which case V_s contains one occurrence of size K. The addition of elements in $N_{exc}(u, V_s)$ along with the $u > v$, ensure that there is no subgraph enumerated twice, and it can be proved [25] that this procedure stops and enumerates all subgraphs.

The tree data structure: During the enumeration process, this data structure is used to encapsulate information about the subgraph contained in V_s. Since this is a recursive procedure, one can use information about the initial content of V_s to build a partial isomorphism representation, that can be complemented on each vertex insertion in V_s. For this, a data structure called a gtrie is used, which is similar to a prefix tree of subgraphs. Whenever a new vertex is added to V_s, one uses the information of connectivity with the previous elements of V_s to generate a label that identifies the current partial subgraph, which is used as the identifier for the mentioned intermediate classes.

Figure 1 summarizes the whole algorithm. The tree on the left represents the implicit recursion tree ESU creates. The induced g-trie on the right is a visual representation of the actual g-trie `FaSE` creates. More information about the `FaSE` can be found in [17].

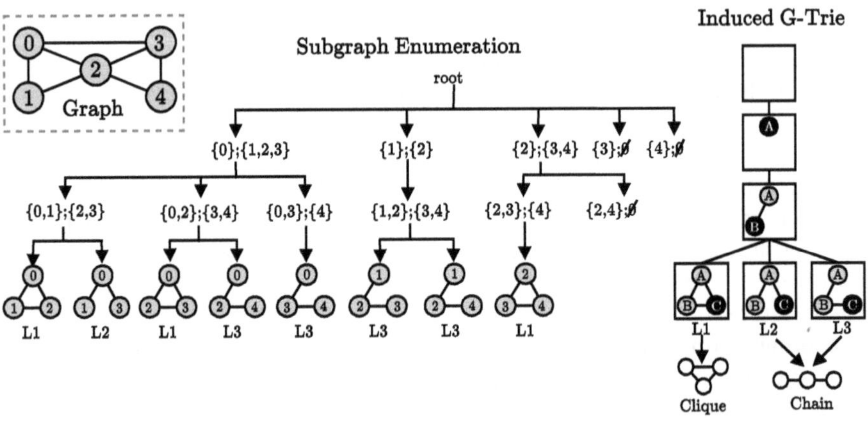

Fig. 1 Summary of the `FaSE` algorithm

4.2 FaSE with Updates

Our method to efficiently update frequency counts works by altering the enumeration algorithm to count frequencies starting on edges. When adding an edge, the algorithm first counts all subgraphs that use the edge's two ends and decrements their frequency. Afterwards, it adds the new edge and counts all subgraphs that touch that edge. To remove an edge we do an analogous process. Our method is based on the ESU algorithm, altering it to start on a given edge.

For a given edge to add, $\{a, b\}$, the algorithm first considers as initial sets $V_s = \{a, b\}$ and $V_{ext} = N(a) \cup N(b) \setminus \{a, b\}$ and only uses these as initial sets (meaning it does not recurse on other initial V_s and V_{ext}). The rest of the procedure is similar to the original ESU algorithm, but the symmetry breaking is removed, that is, when adding a node u' to V_{ext}, there is no comparison with a: if u' belongs to $N_{exc}(u, V_s)$ it will be added to V_{ext}.

To prove that this method is correct we use the original correction proof of the ESU algorithm. If a is the minimal node of the graph (that is, for every node v, $a \leq v$), all subgraphs that include a will be enumerated on the first iteration of the algorithm. For that iteration, if b is the first element of N_{ext}, then it will be removed and the next iteration has $V_s = \{a, b\}$ and $V_{ext} = N(a) \setminus \{b\} \cup N_{exc}(b, \{a\}) = N(a) \cup N(b) \setminus \{a, b\}$. Since this is the only recursion path that will include a and b (since b was the first node to be removed from the initial N_{ext}), all subgraphs that contain a and b will be counted on this recursive subtree. Since this is analogous to our method, its correctness implies the correctness of our method.

5 Experimental Evaluation

We apply our techniques to four networks, two of them neurobiological, based on [23]. The neurobiological networks are directed and represent a macaque visual cortex, with 30 nodes and 311 connections, and a macaque cortex, with 71 nodes and 746 edges. The other two networks are undirected and represent a social network of jazz musicians [7], with 198 nodes and 2742 edges, and a geo-spacial network of a power grid in the United States [24], with 4941 nodes and 6594 edges.

We measure the significance of subgraphs of size $K = 4$ and $K = 5$, using the Z-score metric. For each network and each type of initial random network, we generate an ensemble of 100 random networks. For the convergence phase, we define our energy threshold as 5%, if the vectors of subgraphs count differ in 5% or less, we stop the computation and output the network as it is at that point. We use an initial temperature of 0.01 and a cooling factor of 0.99. Table 1 presents results for the mentioned networks, by comparing the Z-score calculated by our methods against simply maintaining the degree sequence.

Using our generator as null model, the Z-score of the first and second subgraphs on the macaque cortex and fourth, fifth, seventh and eighth on the macaque visual

Table 1 Z-score results for some subgraphs in the macaque cortex and macaque visual cortex networks.

Network	K	Subgraph	Original	Keep $K-1$		
				Keep Deg. Seq.	Change Deg. Seq.	ER
Macaque cortex	4	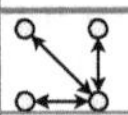	61.20[a]	−2.29	−0.71	−4.41
		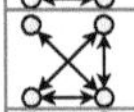	182.30[a]	6.19	2.47	12.66
		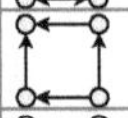	−10.17[b]	12.01	10.64	15.20
Macaque visual cortex	4	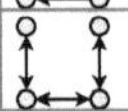	36.76[a]	−1.58	−0.63	−2.88
		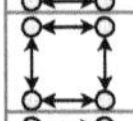	14.63[a]	−2.29	−2.20	−2.61
		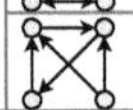	−3.49[b]	12.01	4.90	5.40
	5	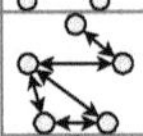	278.57[b]	4.11	3.85	−0.71
		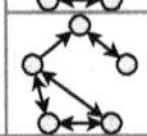	117.72[b]	8.79	6.41	1.62
Power	5	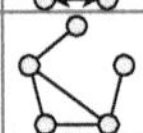	82.83[b]	4.88	−3.45	2.86
		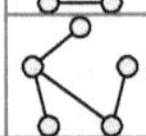	−21.57[b]	−18.25	−17.65	0.09
Jazz	5	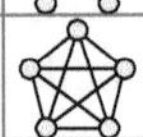	438.35[b]	60.47	29.62	15.82
		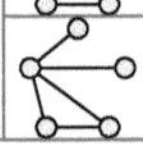	−45.84[b]	−17.31	6.18	70.54

[a]result was taken from [23]. [b]was calculated by us, using degree sequence invariance as null model

cortex was significantly lower than the Z-score calculated using solely the degree sequence as invariant. We speculate that these subgraphs, which are considered over-represented in the original network by Sporns et al. [23], are simply a consequence of the prevalence of their induced subgraphs of size $K-1$. By preserving the frequency of the latter, the former become more common in the generated random networks.

On the other hand, subgraphs third and sixth from macaque cortex and macaque visual cortex respectively, are originally considered under-represented but, under our generator, can be considered motifs. Note that the Z-score values are similar using different initial perturbations on the original networks.

On the `power` network, we show a subgraph of size 5 that was considered a motif under the previous model, but with our new model, it is not considered over-represented anymore. The other example for the same network, using a Markov chain edge swap as the initial network yields a similar Z-score as the original model, but converging from an ER network produces a significantly different score.

For the `jazz` network, we present an example where an extremely over-represented subgraph is still considered a motif under our model. It is the size 5

Table 2 Average execution time, in seconds, and speedup, of the efficient update in comparison with the full census, to generate a random network preserving the frequency of subgraphs of size 3 for the neurobiological networks and size 4 for the `jazz` and `power` networks

	Macaque cortex	Macaque visual cortex	Power	Jazz
Efficient update (s)	64.85	0.22	239.56	1034.06
Full census (s)	103.58	12.35	4274.47	25102.0
Speedup (×faster)	1.6	56.1	17.8	24.3

clique and its over-representation can not be simply explained by the number of size 4 cliques. In the other example, each of the models for the initial random network provides a substantially different Z-score, from being considered under represented if the Markov chain edge swap process that retains the degree sequence is used, to being treated as motif if the initial network follows the ER model.

We also study the improvement obtained by efficiently updating subgraph counts. To this end, Table 2 shows the average execution time, in seconds, for each network, comparing the efficient update against running a full census after each edge swap. These tests were run with initial temperature 0.01, cooling factor set to 0.99 and using the Markov chain edge swap variant that preserves the degree sequence. Subgraph frequency of size 3 was maintained for the macaque networks and size 4 for the `power` and `jazz` networks.

For the macaque cortex network, in average, each network took nearly twice as much doing the full census after each edge switch than using our efficient frequency update. However, for the `jazz` and `power` networks, in average, each network was 1 order of magnitude faster using the efficient update technique and the macaque visual cortex was about 2 orders of magnitude faster.

Clearly, both macaque networks are outliers of efficiency, probably because they are both small dense networks. Our efficient update method works best for larger sparse networks, because in this case, on average, the number of subgraphs that change after a single edge addition or removal is only a small fraction of the total number of subgraphs. In this sense, the `jazz` and `power` networks are better fits for this model, as are most social networks.

6 Conclusion

We introduced a generator of random graphs that preserves the frequency of subgraphs of size $K - 1$. The generation is split in two phases, where the original networks first suffers an initial perturbation, via a Markov chain edge swapping technique or a classic Erdos-Renyi model, and then converges to the desired frequency up to a difference of percentage threshold, using simulated annealing.

We applied our generator to four real complex networks and compared the significance of different subgraphs against results published in [23]. The Z-score calculated by using our generator as null model is significantly lower for certain subgraphs of size K, which can be explained by the prevalence of induced subgraphs of size $K - 1$.

We also devised a technique to efficiently update the frequency of subgraphs after an addition or removal of a single edge. In summary, it works by searching all the subgraphs that touch the edge's endpoints and updates their frequency. This technique is critical to the convergence phase of our generator, as it is, on average, at least 2 times faster and in many cases orders of magnitude faster than running the full networks census from scratch.

Acknowledgements This work is funded within FourEyes, a research line within project *TEC4 Growth/NORTE-01-0145-FEDER-000020.*

References

1. Albert, R., Barabasi, A.L.: Statistical mechanics of complex networks. Rev. Mod. Phys. **74**, 47–97 (2002)
2. Barabási, A.L., Albert, R.: Emergence of scaling in random networks. Science **286**(5439), 509–512 (1999)
3. Bois, F.Y., Gayraud, G.: Probabilistic generation of random networks taking into account information on motifs occurrence. J. Comput. Biol. **22**(1), 25–36 (2015)
4. Choobdar, S., Ribeiro, P., Bugla, S., Silva, F.: Comparison of co-authorship networks across scientific fields using motifs. In: 2012 IEEE/ACM International Conference on ASONAM, pp. 147–152. IEEE (2012)
5. Erdos, P., Rényi, A.: On the evolution of random graphs. Bull. Inst. Int. Stat. **38**(4), 343–347 (1961)
6. Erdos, P., Rényi, A.: On random graphs i. Publ. Math. Debrecen **6**, 290–297 (1959)
7. Gleiser, P.M., Danon, L.: Community structure in jazz. Adv. Complex Syst. **6**(04), 565–573 (2003)
8. Grochow, J.A., Kellis, M.: Network motif discovery using subgraph enumeration and symmetry-breaking. In: Annual International Conference on Research in Computational Molecular Biology, pp. 92–106. Springer (2007)
9. Khakabimamaghani, S., Sharafuddin, I., Dichter, N., Koch, I., Masoudi-Nejad, A.: Quatexelero: an accelerated exact network motif detection algorithm. PloS One **8**(7), e68073 (2013)
10. Kirkpatrick, S., Gelatt, C.D., Vecchi, M.P.: Optimization by simulated annealing. Science **220**(4598), 671–680 (1983)
11. McKay, B.D., Piperno, A.: Practical graph isomorphism, ii. J. Symbo. Comput. **60**, 94–112 (2014)
12. Meira, L.A., Maximo, V.R., Fazenda, A.L., da Conceicao, A.F.: Accelerated motif detection using combinatorial techniques. In: 2012 Eighth International Conference on SITIS, pp. 744–753. IEEE (2012)
13. Milo, R., Itzkovitz, S., Kashtan, N., Levitt, R., Shen-Orr, S., Ayzenshtat, I., Sheffer, M., Alon, U.: Superfamilies of evolved and designed networks. Science **303**(5663), 1538–1542 (2004)
14. Milo, R., Kashtan, N., Itzkovitz, S., Newman, M.E., Alon, U.: On the uniform generation of random graphs with prescribed degree sequences. arXiv preprint arXiv:cond-mat/0312028 (2003)
15. Milo, R., Shen-Orr, S., Itzkovitz, S., Kashtan, N., Chklovskii, D., Alon, U.: Network motifs: simple building blocks of complex networks. Science **298**(5594), 824–827 (2002)
16. Molloy, M., Reed, B.: A critical point for random graphs with a given degree sequence. Random Struct. Algorithms **6**(2–3), 161–180 (1995)

17. Paredes, P., Ribeiro, P.: Towards a faster network-centric subgraph census. In: IEEE/ACM International Conference on ASONAM, pp. 264–271. IEEE (2013)
18. Pinar, A., Seshadhri, C., Vishal, V.: Escape: efficiently counting all 5-vertex subgraphs. arXiv preprint arXiv:1610.09411 (2016)
19. Rao, A.R., Jana, R., Bandyopadhyay, S.: A markov chain monte carlo method for generating random (0, 1)-matrices with given marginals. Sankhyā: The Indian Journal of Statistics, Series A pp. 225–242 (1996)
20. Ribeiro, P., Silva, F.: G-tries: a data structure for storing and finding subgraphs. Data Min. Knowl. Disc. **28**(2), 337–377 (2014)
21. Ritchie, M., Berthouze, L., Kiss, I.Z.: Generation and analysis of networks with a prescribed degree sequence and subgraph family: higher-order structure matters. J. Complex Netwl. cnw011 (2016)
22. Shen-Orr, S.S., Milo, R., Mangan, S., Alon, U.: Network motifs in the transcriptional regulation network of escherichia coli. Nat. Genet. **31**(1), 64–68 (2002)
23. Sporns, O., Kötter, R.: Motifs in brain networks. PLoS Biol **2**(11), e369 (2004)
24. Watts, D., Strogatz, S.: Collective dynamics of small-world networks. Nature **393**, 440–442 (1998)
25. Wernicke, S.: Efficient detection of network motifs. IEEE/ACM Trans. Comput. Biol. Bioinform. (TCBB) **3**(4), 347–359 (2006)

Fuzzy Centrality Evaluation in Complex and Multiplex Networks

Sude Tavassoli and Katharina A. Zweig

Abstract Centrality rankings are classically used to analyze the influence of nodes in different types of networks. However, since most centrality indices are very sensitive to missing or additional edges and since most complex networks are based on faulty data, a precise ranking is quite unlikely to be obtained. Thus, in this paper we propose to use an assignment of the nodes to a predefined and small set of centrality classes using a fuzzy model, ranging from "very peripheral" to "very central". We show empirically that the assignment of nodes to these classes is quite robust against random noise. Furthermore, the method can also be used to combine possibly conflicting classes of the nodes based on different centrality values over multiple networks using a fuzzy operator.

1 Introduction

Many real networks are based on incomplete data that demands new network analytic approaches which can handle uncertainty issues. For example, the number of connections that a node has in a network, might not be exactly the one that is logged in the dataset. Then, analyzing centrality indices and obtaining a precise ranking might underestimate the existing uncertainty. One way of dealing with such issues is the use of fuzzy models in the corresponding analysis. In this paper, we consider the analysis of normalized degree and closeness centrality in multiplex networks as a decision making problem and aim at analyzing them within all the layers using fuzzy logic models. It has been shown in many studies that fuzzy models can deal with the issues coming from uncertainty and can avoid information loss in decision making problems [10–12, 18–20]. Therefore, in a wide range of studies from different fields,

S. Tavassoli (✉) · K. Zweig
Graph Theory and Complex Network Analysis Group, Computer Science Department, Kaiserslautern University of Technology, Gottlieb-Daimler-Str. 48, 67663 Kaiserslautern, Germany
e-mail: tavassoli@cs.uni-kl.de

K. Zweig
e-mail: zweig@cs.uni-kl.de

B. Gonçalves et al. (eds.), *Complex Networks VIII*,
Springer Proceedings in Complexity, DOI 10.1007/978-3-319-54241-6_3

these models have been used, such as expressing medical situations [5], analyzing ranking methods [2], fuzzy rule-based classification systems [6], and analyzing central nodes in fuzzy cognitive maps [16]. In the last study [16], a fuzzy linguistic model (proposed by Herrera in [10]) is used to obtain the central nodes by using several classical centrality measures in fuzzy cognitive maps, which is a model for representing a domain knowledge and the connections between different factors of the domain. Our work is similar to that study in the sense of using the fuzzy model for proposing a new centrality concept but in a different research area.

We aim at addressing the question to which degree a node is central instead of its seemingly precise ranking position in multiple layers of a multiplex network. As always when a node's centrality needs to be compared over multiple networks, it is necessary to normalize the values beforehand. Furthermore, there are different strategies of aggregation. We have shown that the sensitivity of rankings—obtained from only the degree centrality—to the choices of different modeling decisions in multiplex networks, can heavily influence the findings and thus the interpretations [21]. This encourages us to find a solution for dealing with the sensitivity of rankings, especially in multiplex representations, where multiple types of interaction have different roles in the identification of influential nodes. We present the results using new visualizations that show the assignments of the nodes to a set of predefined classes of fuzzy centrality.

The rest of this paper is organized as follows: Sect. 2 discusses the motivation and the used dataset. Section 3 elaborates the theoretical background of multiplex networks and fuzzy logic models. Section 4 explains the steps of fuzzy centrality evaluation and Sect. 5 contains all the experimental results including the visualizations. Finally, Sect. 6 summarizes the study.

2 Motivations and the Used Dataset

In most of the studies related to centrality concept—from simple graphs to multiplex networks—a centrality ranking is mostly used to present the importance of the nodes with respect to a centrality measure. In our recent study [21], we have shown that very basic and seemingly simple modeling decisions like different normalizations and aggregations used for the degree centrality in a multiplex network, can change a node's ranking from being the most central to the least central. This is because the different choices of modeling decisions result in conflicting rankings of the nodes and this changes the interpretations of the findings. Next to the mentioned problem in a multiplex network, some studies have also shown the effects of missing data in different types of real networks on their analysis regarding the centrality concept [3, 15]. By contrast, some other studies have explained that centrality measures are almost robust to random network errors and thus, finding the confidence interval around a centrality index value is not impossible [1, 4]. These different views about centrality measures motivated us to find a model that can facilitate the evaluation of centrality in many real complex networks. Therefore, in this paper, Centrality is considered as a fuzzy concept and instead of giving the nodes a discrete and exact ranking position with respect to their normalized centrality values in a multiplex network, the nodes are partitioned into groups of about the same centrality. The robustness of the model to random edge deletion and edge addition is then represented at the end of the evaluation.

We use a real network containing multiple interactions between 79 individuals in the so-called **Noordin terrorist group**, which was behind several terrorist attacks from 2003 to 2005. The data was drawn from a report in 2006 [14] and was then structured by Roberts et al. [17]. It was also analyzed in detail as a so-called "dark network" by Everton and Cunningham [7]. The dataset encompasses very rich information about different types of relations and interactions among the members as well as their attributes such as military training, nationality, and their education level. In this paper, we use three different types of interactions as three layers of a multiplex network as follows: the *trust network*, which is the aggregated version of four different ties representing friendship, classmate, soulmate, and kinship relations among the 79 members. The *operational network* aggregates relations that exist if two individuals provided the same logistics, were in common meetings, participated in common operations, or in the same training event. The *communication network* represents whether two individuals communicated using messages inside the group or had a communication using external mediums such as codes and videos to recruit the other members outside the group. All single layers are represented as simple graphs where two nodes are connected if the corresponding persons are in at least one of the aforementioned relations in the respective layer.

3 Theoretical Background

3.1 Multiplex Networks and Centrality Measures

In this section, all the preliminaries required for the fuzzy centrality evaluation in multiplex networks are described in detail.

Definition 1 A multiplex network is a network comprised of $|L| = n$ layers $L = \{l_1, l_2, \ldots, l_n\}$, where each layer l_i is a simple graph with a set of nodes denoted by V_i and a set of edges $E_i \subseteq V_i \times V_i$, which represents a specific type of interaction in the layer l_i.

The centrality measures in a simple graph are defined by Freeman [9] and here are extended to a multiplex network.

Definition 2 The degree of a node v is the number of edges connected to the node v in layer l_i and denoted by $deg_i(v)$. Then, the normalized degree of the node v in layer l_i is obtained by

$$C_D(v, l_i) := \frac{deg_i(v) - min\{deg_i(v)|v \in V_i\}}{max\{deg_i(v)|v \in V_i\} - min\{deg_i(v)|v \in V_i\}}, \quad (1)$$

Definition 3 Let $d_i(v, u)$ denote the distance of two nodes in layer l_i which is defined if and only if $v, u \in V_i$. The closeness of a node v is the inverse of the sum of its distances to all the other nodes in the largest connected component of layer l_i.

$$close_i(v) = \left[\sum_{u \neq v} d_i(v, u)\right]^{-1}, \quad (2)$$

Accordingly, the normalized closeness centrality $C_C(v, l_i)$ of the node v in layer l_i can be obtained as measured for the degree centrality.

3.2 *Fuzzy Logic Models*

Fuzzy linguistic models were introduced by Zadeh [22] and further used in a fuzzy 2-tuple model introduced by Herrera [10] to solve a decision making problem.

Definition 4 A Multi-Criteria Decision Making problem searches for a satisfying solution among a set of alternatives based on multiple, conflicting criteria.

In a way, searching for the most central nodes in a multiplex network is such an *MCDM* problem: while some nodes are perfectly central in some of the layers and rather peripheral in other layers, other nodes might be quite central in all layers, but only a few nodes are perfectly central in all. Now, an *MCDM* problem searches for a solution (a node) that fulfills the criteria (high centrality) the best.

In earlier work [21], we either directly used the normalized centrality values or we used the ranking position of the nodes. In this paper, based on the observation that incomplete data might change the ranking, we assign each node in each layer in one of five categories, from "very peripheral" to "very central". The five categories and their textual description form a *linguistic term set*. For obtaining the category of centrality for the node v in the layer l_i, we fuzzify its normalized centrality index in the interval of [0, 1] by Gaussian membership functions, each of which describes a term in a linguistic term set. Herrera explains that each value in this interval can be fuzzified using these functions [11].

For the fuzzification, assume that $S = \{s_0, \ldots, s_g\}$ is a linguistic term set and $s_i \in S$ is a linguistic term. Each term s_i can then be described using a symmetric Gaussian function including some parameters, which can be obtained using a fuzzy toolbox. Given a set of five classes (labels), and an ordering of these classes as shown in Fig. 1, the membership values μ_{s_i} of the normalized centrality $C_X(v)$ of any node v is then determined by the intersection of the value and the corresponding class of s_i.

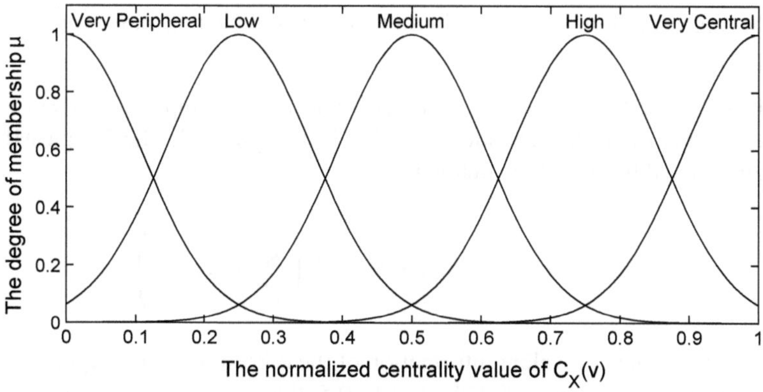

Fig. 1 Five linguistic terms and their semantic are described using five overlapping Gaussian membership functions

A vector that contains the membership values for a node and all classes is called a fuzzy set, as exemplified in Fig. 1 for the value of 0.15. A symbolic aggregation operation can then be used to obtain an aggregated value over the fuzzy set [11].

Definition 5 Let $T = \{(s_0, \mu_{s_0}), (s_1, \mu_{s_1}), \ldots, (s_g, \mu_{s_g})\}$ be a fuzzy set, then a symbolic aggregation operation is as follows:

$$\beta = \frac{\sum_{j=0}^{g} j \cdot \mu_{s_j}}{\sum_{j=0}^{g} \mu_{s_j}} \tag{3}$$

The result of the aggregation is $\beta \in [0, g]$, i.e., it is a linear projection onto the sequence of the linguistic term set.

Definition 6 Let β be the result of symbolic aggregation, then the equivalent information of β in the linguistic term set S can be expressed using a 2-tuple model by the function of $\Delta : [0, g] \rightarrow S \times [-0.5, 0.5)$:

$$\Delta(\beta) = (s_i, \alpha), \ with \begin{cases} s_i, & i = round(\beta) \\ \alpha = \beta - i, & \alpha \in [-0.5, 0.5), \end{cases} \tag{4}$$

where *round* is the usual operation of rounding as defined by Herrera in [11]. That is, the 2-tuple contains the category of which a node is mostly the member of, plus the parameter $\alpha \in [-0.5, 0.5)$, which is the value of the symbolic translation. It supports the *difference of information* between β and the closest index to it in $\{0, \ldots, g\}$.

In our last work [21], we used the Maximum Entropy Ordered Weighted Average introduced by Yager to deal with the conflicting criteria in an MCDM problem [8, 21]. The MEOWA fuzzy operator including the 2-tuples is defined by Herrera to deal with the aggregation of multiple criteria that their satisfactions are expressed using a 2-tuple [10, 11].

Definition 7 Let $A = \{(a_1, \alpha_1), \ldots, (a_n, \alpha_n)\}$ be a set of 2-tuples and $w = (w_1, \ldots, w_n)$ be a weight vector that satisfies $w_i \in [0, 1]$ and $\sum w_i = 1$. The 2-tuple MEOWA operator denoted by F^e is then defined:

$$F^e((a_1, \alpha_1), \ldots, (a_n, \alpha_n)) = \Delta\left(\sum_{j=1}^{n} w_j \ldots \beta_j^*\right), \quad (5)$$

where each β is the equivalent information of the corresponding 2-tuple and can be obtained using the function of $\Delta^{-1} : S \times [-0.5, 0.5) \to [0, g]$ as follows:

$$\Delta^{-1}(s_i, \alpha) = i + \alpha = \beta, \quad (6)$$

Assume that all the β-values are obtained, then the operator multiplies the weight vector by the non-increasingly sorted version of the vector of β-values (denoted by β^*). Then, it again uses the Δ-function to result in a 2-tuple. For obtaining the MEOWA weight vector, the following function based on a parameter γ is introduced by Yager [8]:

$$w_i = \frac{e^{\gamma \frac{n-i}{n-1}}}{\sum_{j=1}^{n} e^{\gamma \frac{n-j}{n-1}}}, \quad (7)$$

For $\lim \gamma = \infty$, the weight vector is $(1, 0, \ldots, 0)$. Thats it, since the classes are sorted, the aggregation strategy returns the maximum value. In total, such an aggregation operation will select as the most central node with the best class of centrality in *any layer*. For $\lim \gamma = -\infty$, the weight vector is $(0, 0, \ldots, 1)$ and the minimum value has the most important role in the aggregation strategy, which means a node with the highest class of centrality in *all layers* is of favor. When $\gamma = 0$, the weight vector is $(\frac{1}{n}, \frac{1}{n}, \ldots, \frac{1}{n})$ and the aggregation strategy is a regular *average* over the β-values.

4 Fuzzy Centrality Evaluation

The proposed evaluation model is comprised of multiple steps. First, a normalized centrality index value of a node is fuzzified using the membership functions and a fuzzy set including the classes and the membership values are obtained. Second, an aggregated value over the fuzzy set is computed, whose equivalent information in the predefined term set including five classes, is expressed using a fuzzy 2-tuple—this results in a class of fuzzy centrality for a node and the degree to which the node is close to its class in a network layer. Then, having all centrality classes for the node over multiple layer obtained, the question to be addressed is to which class of centrality the node can be assigned with respect to different aggregation strategy. Therefore, the next step is dedicated to the usage of the MEOWA operator to produce the different aggregations. Again, the fuzzy 2-tuple model is used to determine the class of centrality after the aggregation. Finally we show the robustness of the model with respect to noises. We use randomly edge deletion and edge addition with the rates of 10, 20, and 30%$|E_i|$. For adding a single edge, a pair of nodes is randomly selected and checked whether they are not connected but have a common neighbor, then the edge is added. For each rate of noise, we perform the procedure 50-times and obtain an average over all the obtained closeness centrality values for a node. For removing the edges, a sequence of edges are randomly selected and removed from a layer, then the closeness centrality of the nodes is measured in the largest connected component in the corresponding layer. Afterward, the differences are measured between the assignments of classes of centrality for all nodes in the original network layer and the layer including the noise using the following equation:

$$d_{class}(l_i, l_i') := \frac{\sum_{v=1}^{|V_i|} |(s, \alpha)_v - (s', \alpha')_v|}{g \ldots |V_i|}, \tag{8}$$

where $(s, \alpha)_v$ is the class of centrality of a node v in the layer l_i and $(s', \alpha')_v$ is its class of centrality in the layer with the noise l_i' and g is the maximum change that a node can have within the five classes, which is 4 here. Note that using Eq. (6), the $\beta-$values of the classes can be achieved. The lower and upper bound of d_{class} is $[0, 1]$. A value close to 0 indicates that overall, the nodes have minimum changes in their fuzzy centrality classes between the original layer and the layer with the noise and a value close to 1 indicates that the nodes have maximum changes in between.

5 Experimental Results

All the nodes in the three layers of *trust network, operational network,* and *communication network* are colored in Fig. 2 with respect to their classes based on degree centrality. The isolated nodes in the layers are removed from the visualization. Among the nodes in the *trust network*, A. Sungkar in the center of the network is in the highest class of centrality, which is labeled **Very Central**.

The five nodes around him in the class of **High** degree centrality, are Noordin, M. Rais, Tohir, A. B. Ba'asyir, and F. Al. Ghozi. Noordin is always in the best class in the layers of *operational network* and *communication network*. In terms of doing the operations, A. Husin with the role of Bomb expert is the second node that is assigned to the class of **Very Central**. A. Dharmawan as the coordinator of attack and logistics is the only one in the class of **High** centrality in the layer of *operational network*. In contrast to the other layers, the nodes in the layer of *communication network* are mainly assigned to the classes of **Very Peripheral** and **Low** centrality.

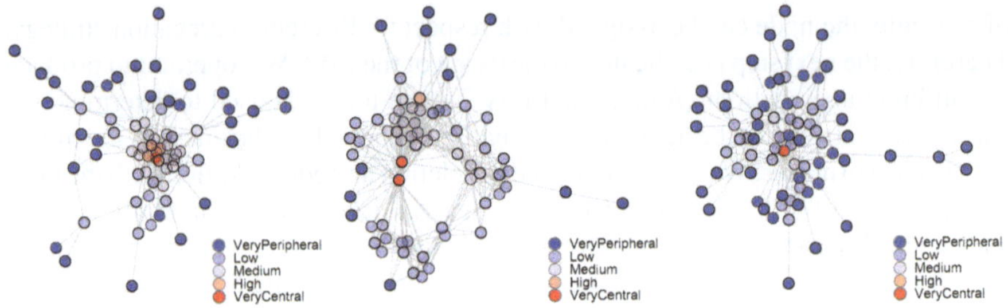

(a) the *trust network* (b) the *operational network* (c) the *communication network*

Fig. 2 The assignments of the nodes to the five classes of degree centrality are separately demonstrated for the three layers of the network

For the multiplex evaluation, as visualized in Fig. 3, the nodes are assigned to all of the five previously defined classes of centrality, i.e., the classes in the x-axis and the $|\alpha|$-value in the y-axis. In order to avoid the overlapping of the nodes' label, we use the α-value in the x-axis as well. This makes the node that has a negative α-value, stay before the middle line of its label (ticked) and the one that has a positive α-value, shift ahead of the label. Note that in this visualization, the nodes can have the same class and the same $|\alpha| - value$, thus a point can be dedicated to several nodes.

The importance of the nodes with respect to their highest class of centrality in *at least one* or in *all the layers* is shown in Fig. 3. In the first aggregation strategy, where a node with the most satisfying classes of centrality in *all the layers*, is identified as the most important one, four terrorists are the most distinguished, key members in the organization with the different classes of centrality: A. R. Ridho who is the Noordin's courier, A. Dharmawan as the coordinator of attack and logistics, A. Husin as the Bomb expert, and Noordin on the top. In a recently published study [13], it has been shown that using several types of multiplex page rank, these four members are always on top—which is the same result as observed here. A. Sungkar, who had the role of strategist in the dataset, is not among the top nodes in the first aggregation strategy as he has only one role with a satisfying class of centrality, not in *all the layers*. However, when having *at least one* important role is enough to identify a node as the important one, he is among the **Very Central** nodes as shown in Fig. 3b, because of his important role in the layer of *trust network*. There are many cases that are assigned to the classes of **Medium** and **High** in the second aggregation strategy, which were rather not distinguished as important nodes in the first aggregation; this means their least class of centrality was not satisfying enough to represent them as the key members.

For the fuzzy closeness centrality, we used the similar visualization, as shown in Fig. 4. All the detailed visualizations including the classes and $|\alpha| - value$ are depicted for the three layers in Fig. 5a–c respectively. The four nodes: Ubeid (who was jailed after four months being in the organization), A. Sungkar, Noordin and A. R. Ridho are categorized into the class of **Very Central** in the layer of *trust network*. Although the first node did not have a direct connection with Noordin in this layer,

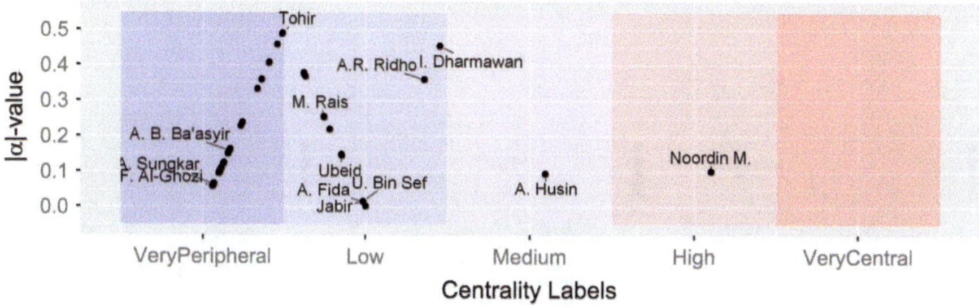

(a) all layers

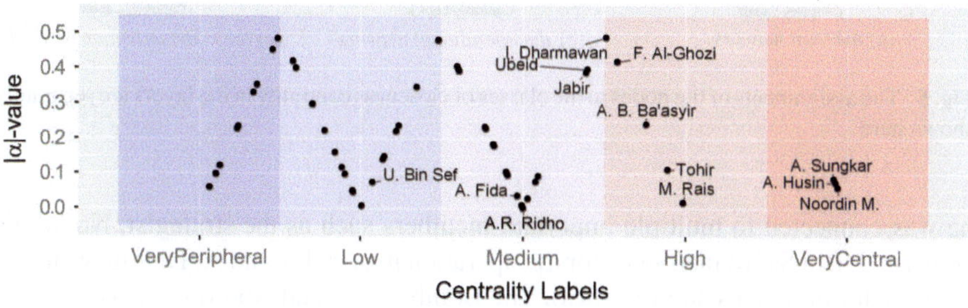

(b) at least one

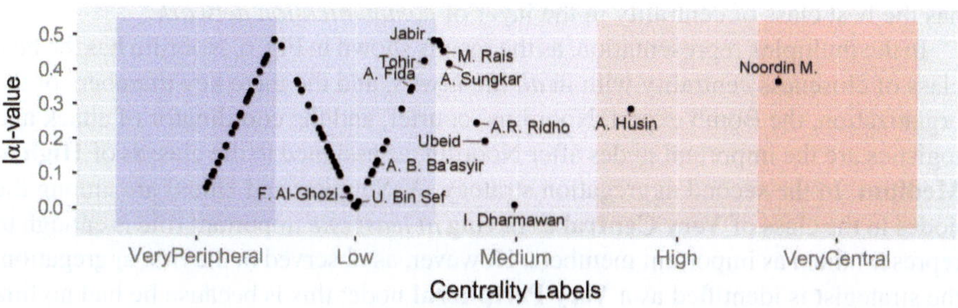

(c) on average

Fig. 3 The assignments of 79 individuals to the classes of fuzzy degree centrality are depicted based on the results of the different aggregations over the three layers

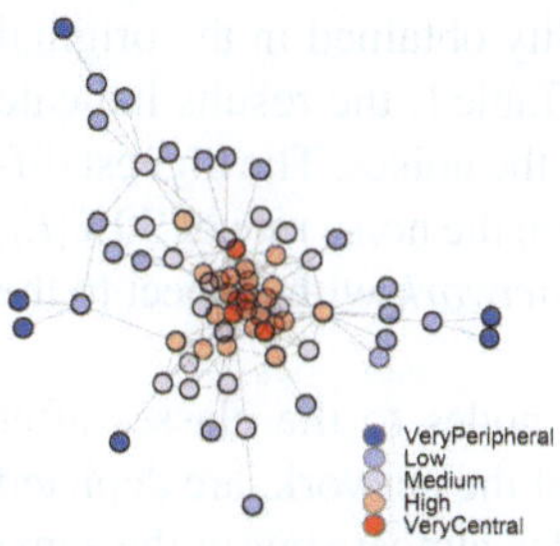

(a) the *trust network*

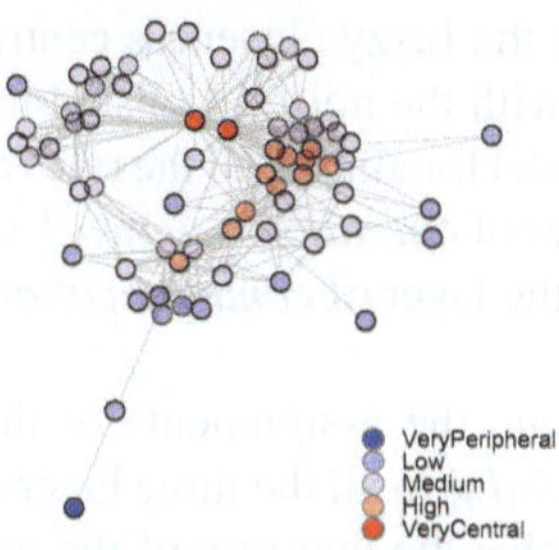

(b) the *operational network*

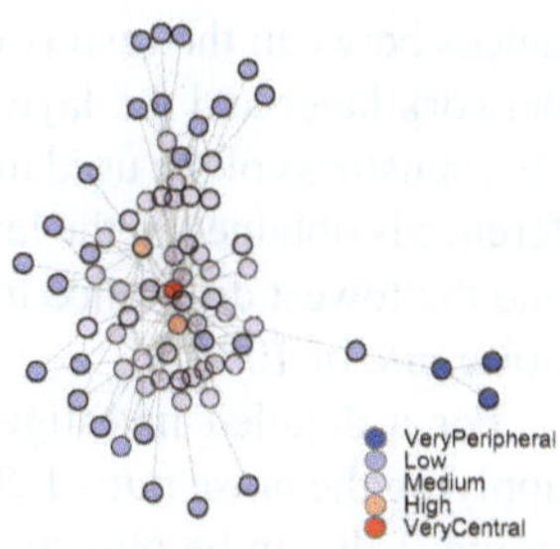

(c) the *communication network*

Fig. 4 The assignments of the nodes to the five classes of closeness centrality are separately demonstrated for the three layers of the network

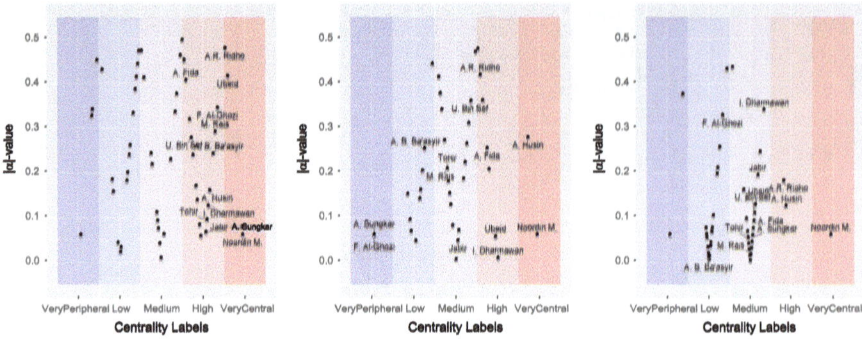

(a) *the trust network* (b) *the operational network* (c) *the communication network*

Fig. 5 The assignments of the nodes to the classes of closeness centrality in the layers are separately shown here

he was connected to multiple important members such as the strategist, Noordin's courier, U. Bin Sef (a facilitator for the operation materials), and A. Dharmawan (the coordinator of attacks and logistics). The Bomb expert and Noordin are assigned to the class of **Very Central** in the layer of *operational network* and Noordin top also has the best class of centrality in the layer of *communication network.*

In the multiplex representation, as the results shown in Fig. 6, Noordin has the best class of closeness centrality with in *all the layers*, and the three key members of the organization, the Bomb expert, Noordins' courier, and the coordinator of attack and logistics are the important nodes after Noordin, as assigned to the classes of **High** to **Medium**. In the second aggregation strategy, A. Sungkar and Ubeid are among the nodes in the class of **Very Central** as having *at least one* important role is enough to represent them as important members. However, as observed in the first aggregation, the strategist is identified as a **Very Peripheral** node; this is because he had no link to others regarding the operations and also had very small communication links to the others in the *communication network.* Thus, when his role in *all the layers* is evaluated, he belongs to the class of **Very Peripheral** nodes.

In order to show the robustness of the model to the noises, we measure the distances between the results of the fuzzy closeness centrality obtained in the original network layer and the layer with the noise. As listed in Table 1, the results indicate the robustness of the used model for almost all the rates of the noises. The highest difference is obtained in the layer of *operational network* with the noise rate of $30\%|E_i|$ and the lowest difference in the layer of *communication network* with respect to the noise rate of $10\%|E_i|$.

For a detailed investigation, the assignments of the nodes to the classes after applying the noise rate of $20\%|E_i|$ to all the three layers of the network, are depicted in Fig. 7. It can be observed that the majority of the nodes almost stay in the same class of centrality in comparison with the obtained results in the original network layers as shown in Fig. 5.

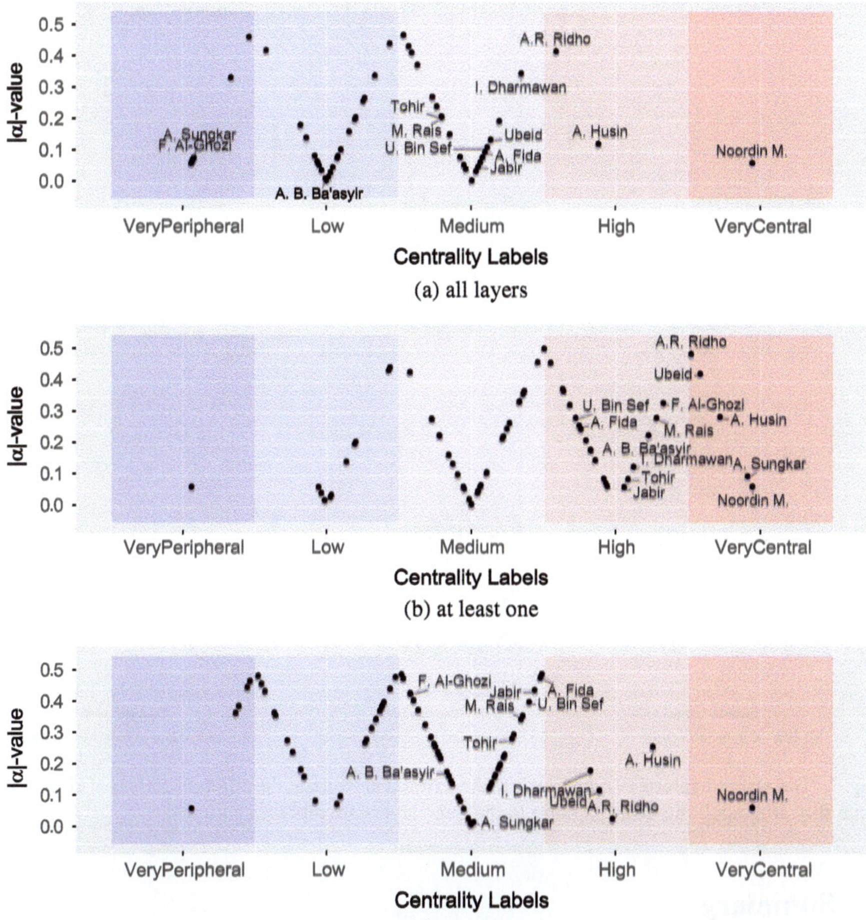

(a) all layers

(b) at least one

(c) on average

Fig. 6 The assignments of the nodes to the classes of closeness centrality after the aggregation of the results over the layers

Table 1 The differences of the assignments of the classes (denoted by $d_{class}(l_i, l'_i)$) between the original network layer of l_i and the layer of l'_i including the noise are listed

	Edge deletion			Edge addition		
	10%	20%	30%	10%	20%	30%
Trust network	0.047	0.061	0.072	0.015	0.028	0.056
Operational network	0.054	0.089	0.106	0.025	0.043	0.059
Communication network	0.041	0.065	0.091	0.006	0.014	0.025

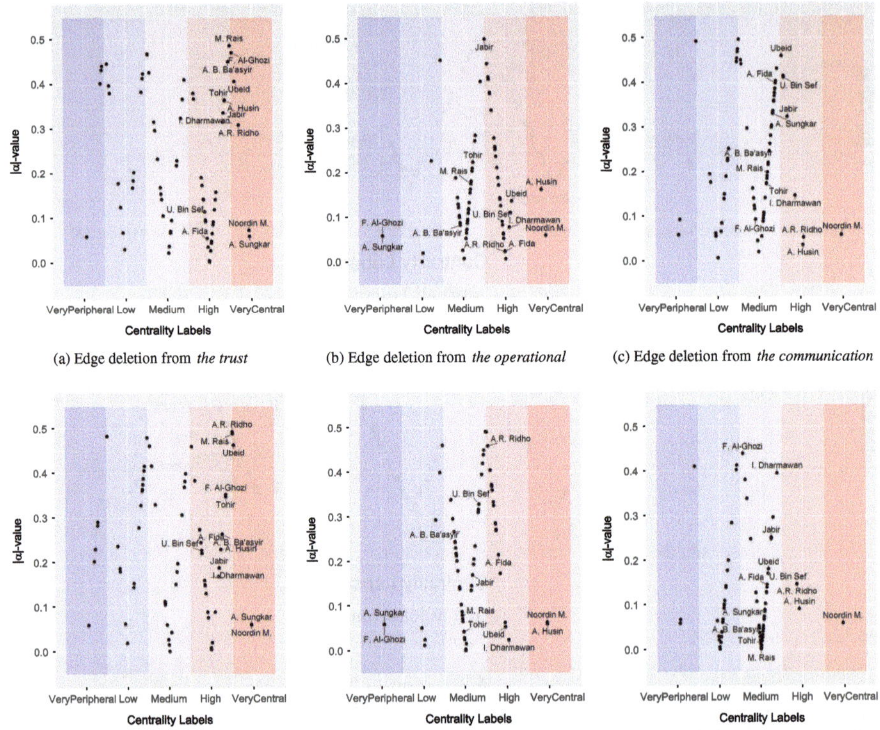

(a) Edge deletion from *the trust* (b) Edge deletion from *the operational* (c) Edge deletion from *the communication*

(d) Edge addition to *the trust* (e) Edge addition to *the operational* (f) Edge addition to *the communication*

Fig. 7 The assignments of 79 individuals in the terrorist network to the different classes of closeness centrality after applying $20\%|E_i|$ noise to the three layers are represented

6 Summary

In this paper we aim at evaluating the centrality of nodes within the layers of a multiplex network with respect to their fuzzified, normalized centrality values. The used fuzzy logic model allows for a comprehensive evaluation of centrality in complex and multiplex networks. Since centrality rankings can be sensitive to the different choices of modeling decisions in multiplex networks or to the incompleteness of a network data, in this model instead of using a discrete ranking, the nodes are partitioned into groups of nodes with the same classes of centrality. In order to show the robustness of the model, several noises are applied to the layers of the network and the differences of the assignments between the original network and the network including the noises are measured. The empirical results show that the model is almost robust to the noises. This model will be even more practical for evaluating the concept of fuzzy centrality, when the classes of centrality are modeled using unbalanced fuzzy partitions.

References

1. Borgatti, S.P., Carley, K.M., Krackhardt, D.: On the robustness of centrality measures under conditions of imperfect data. Soc. Netw. **28**(2), 124–136 (2006)
2. Brunelli, M., Mezei, J.: How different are ranking methods for fuzzy numbers? a numerical study. Int. J. Approximate Reasoning **54**(5), 627–639 (2013)
3. Carrington, P.J., Scott, J., Wasserman, S.: Models and Methods in Social Network Analysis, vol 28. Cambridge university press (2005)
4. Costenbader, E., Valente, T.W.: The stability of centrality measures when networks are sampled. Soc. Netw. **25**(4), 283–307 (2003)
5. Degani, R., Bortolan, G.: The problem of linguistic approximation in clinical decision making. Int. J. Approximate Reasoning **2**(2), 143–162 (1988)
6. Elkano, M., Galar, M., Sanz, J., Bustince, H.: Fuzzy rule-based classification systems for multi-class problems using binary decomposition strategies: on the influence of n-dimensional overlap functions in the fuzzy reasoning method. Inf. Sci. **332**, 94–114 (2016)
7. Everton, S.F., Cunningham, D.: Detecting significant changes in dark networks. Behavior. Sci. Terror. Polit. Aggress. **5**(2), 94–114 (2013)
8. Filev, D., Yager, R.R.: Analytic properties of maximum entropy owa operators. Inf. Sci. **85**(1), 11–27 (1995)
9. Freeman, L.: Centrality in social network, conceptual clarification. Soc. Netw. **1**, 215–239 (1979)
10. Herrera, F., Martinez, L.: A 2-tuple fuzzy linguistic representation model for computing with words. IEEE Trans. Fuzzy Syst. **8**(6), 746–752 (2000)
11. Herrera, F., Martinez, L.: A model based on linguistic 2-tuples for dealing with multigranular hierarchical linguistic contexts in multi-expert decision-making. IEEE Trans. Syst. Man Cybern. Part B (Cybern.) **31**(2), 227–234 (2001)
12. Herrera, F., Herrera-Viedma, E., Martinez, L.: A fuzzy linguistic methodology to deal with unbalanced linguistic term sets. IEEE Trans. Fuzzy Syst. **16**(2), 354–370 (2008)
13. Iacovacci, J., Bianconi, G.: (2016) Extracting information from multiplex networks. Chaos Interdisc. J. Nonlinear Sci. **26**(6):065–306
14. Jones, S.: Terrorism in indonesia: Noordins networks. Tech. rep, Asia Report (2006)
15. Kossinets, G.: Effects of missing data in social networks. Soc. Netw. **28**(3), 247–268 (2006)
16. Obiedat, M., Samarasinghe, S., Strickert, G.: A new method for identifying the central nodes in fuzzy cognitive maps using consensus centrality measure (2011)
17. Roberts, N.: Roberts and everton terrorist data: Noordin top terrorist network (subset). Machinereadable data file (2011)
18. Rodriguez, R.M., Martinez, L., Herrera, F.: Hesitant fuzzy linguistic term sets for decision making. IEEE Trans. Fuzzy Syst. **20**(1), 109–119 (2012)
19. Rodriguez, R.M., Martinez, L., Herrera, F.: A group decision making model dealing with comparative linguistic expressions based on hesitant fuzzy linguistic term sets. Inf. Sci. **241**, 28–42 (2013)
20. Rodriguez, R.M., Martinez, L., Torra, V., Xu, Z., Herrera, F.: Hesitant fuzzy sets: state of the art and future directions. Int. J. Intell. Syst. **29**(6), 495–524 (2014)
21. Tavassoli, S., Zweig, K.A.: Most central or least central? how much modeling decisions influence a node's centrality ranking in multiplex networks. arXiv preprint arXiv:160605468 (2016)
22. Zadeh, L.A.: Fuzzy sets. Inf. Control **8**(3), 338–353 (1965)

Part II
Community Structure

Enhancing Space-Aware Community Detection Using Degree Constrained Spatial Null Model

Remy Cazabet, Pierre Borgnat and Pablo Jensen

Abstract Null models have many applications on networks, from testing the significance of observations to the conception of algorithms such as community detection. They usually preserve some network properties, such as degree distribution. Recently, some null-models have been proposed for spatial networks, and applied to the community detection problem. In this article, we propose a new null-model adapted to spatial networks, that, unlike previous ones, preserves both the spatial structure and the degrees of nodes. We show the efficacy of this null-model in the community detection case on synthetic networks.

1 Introduction

In recent years, complex networks have become an important topic of research, and are used to model systems and interactions in many different fields, from social sciences to biology.

When elements represented as vertices have a location in space, and the distance between them plays a role, we use *spatial networks* to represent them. Examples of networks modelled by spatial networks include transportation networks, infrastructure networks, mobility networks, or even neural networks. Several models of spatial networks exist, such as random planar graph [1], or generalizations of the Watts-Strogatz model. The distinctive characteristic of spatial network models is that the probability of observing an edge between vertices depends on the distance between them. This characteristic can be represented by a *deterrence function*. For a broad overview of existing work on spatial networks, one can turn to [2].

R. Cazabet (✉)
Sorbonne Universites, UPMC Univ Paris 06, CNRS, LIP6 UMR 7606, Paris, France
e-mail: remy.cazabet@gmail.com

P. Borgnat · P. Jensen
CNRS, Laboratoire de Physique, Univ Lyon, Ens de Lyon, Univ Claude Bernard, Villeurbanne, France

B. Gonçalves et al. (eds.), *Complex Networks VIII*,
Springer Proceedings in Complexity, DOI 10.1007/978-3-319-54241-6_4

In complex networks, *null-models* are frequently used to compare the observed properties (assortativity, diffusion, clustering, frequency of patterns, etc.) of a collected network with the ones in a randomized version of it. Another common usage is in community detection, where a quality function called Modularity compares the fraction of edges found inside communities in the observed network and in the corresponding null-model. The most commonly used null model, often called the configuration model (see Sect. 2.1.1), rewires randomly connections between vertices while conserving the degree distribution.

Previously proposed null-models for spatial networks conserve the position of nodes, the deterrence function and the total number of edges, but not the degree distribution. In this article, we propose a null model for spatial networks that preserves as much as possible both the spatial properties and the degrees of nodes.

1.1 Related Works

In [3], the authors propose a method to find space-independent communities in spatial networks. They successfully uncover a linguistic partition in a Belgian mobile phone calls dataset, that was otherwise hidden by geographical proximities. To do so, they use a modified version of the quality function called Modularity (see Sect. 2.3). We will detail this gravity-based null model in Sect. 2.1.2. They use a modified version of the Louvain algorithm [4] to optimize their variant of modularity. Several articles, for instance [5, 6], applied this approach on different case studies.

In [7], the authors propose to use a mechanism similar than the one in [3], but replace the gravity-based spatial null-model by a radiation-based one. The radiation model has been recently proposed as an alternative to the gravity one, and has attracted a lot of attention since then. Their model is described in Sect. 2.1.3. They do not use the exact same method than [3] to optimize their quality function, but a variant of it.

2 Description of Evaluation Settings

2.1 Description of State-of-the-art Null-Models

2.1.1 Configuration Model

The configuration model, or NG model, has been introduced in [8]. It proposes to rewire randomly the graph while keeping the degrees of nodes.

2.1.2 Gravity-Based

This null-model introduced in [3] is based on works coming from the transportation domain, where gravity models have long been used to model the repartition of trips among areas such as cities, countries or neighborhoods.

In recent works, a general version of the law is often used [9],

$$P_{ij}^{Gra} = n_i n_j f(d_{ij}) \tag{1}$$

with $f(d)$ any deterrence function, and n_i same as before. Instead of being decided a priori, the deterrence function can be learned from the data as follows [7]:

$$f(d) = \frac{\sum_{i,j|d_{ij}=d} A_{ij}}{\sum_{i,j|d_{ij}=d} n_i n_j} \tag{2}$$

with A_{ij} the observed flow (number of trips, communications, etc.) between nodes i and j, and d_{ij}, n_i same as in Eq. 1.

We can note that if the distance has no effect, the deterrence function is a constant function, and the gravity-based model becomes exactly the configuration model.

2.1.3 Radiation-Based

Just as the gravity law is an analogy of Newton's law of gravity, the radiation model takes his inspiration form laws of radiation in physics. It has first been introduced in [10], and has been successfully applied in several cases since. It is defined as:

$$P_{ij}^{Rad} = T_i \frac{n_i n_j}{(n_i + r_{ij})(n_i + n_j + r_{ij})} \tag{3}$$

with $r_{ij} = q_{ij} - (n_i + n_j)$, q_{ij} being the sum of n_k for all k in the circle of center i and radius d_{ij} (population closer from i than j). Other notations identical to Eq. 7.

A particularity of this model is that it does not need an explicit deterrence function, as the interactions between nodes depends on their *intrinsic strength* and of the presence of other nodes around them.

To be able to tune the importance of distance, however, the variant of the radiation model introduced in [7] adds a deterrence function effect learned from data, identical to the one previously introduced. The Distance Tuned radiation model becomes:

$$P_{ij}^{DTRad} = P_{ij}^{rad} f(d_{ij}) \tag{4}$$

with $f(d_{ij})$ a deterrence function defined as in the gravity-based case.

2.2 Synthetic Benchmarks for Space-Corrected Community Detection

The benchmark introduced in [3] in a gravity-based version and extended in [7] to a radiation process generates a network with both a planted community structure and a spatial structure. Its distinctive feature is that all edges probabilities have to respect the spatial structure. Compared with the version presented in [7], we introduce two minor modifications:

- We generalize it in order to allow any deterrence function
- We allow the gravity version to handle variable intrinsic weights

The generic test benchmark is defined as:

$$p_{ij}^{Inc} = \lambda(c_i, c_j) P_{ij}^{SNM}(f(d_{ij}))Z_1 \tag{5}$$

with c_i the community containing node i, the function $\lambda(c_i, c_j) = 1$ if nodes i and j are in the same community, and $\lambda(c_i, c_j) = \lambda_d$ otherwise, $P_{ij}^{SNM}(f(d))$ a probability given by the chosen spatial null model with deterrence function $f(d)$, and Z_1 a normalization constant ensuring that $\Sigma_{i>j} p_{ij}^{Inc} = 1$.

Parameters are: N the number of nodes, C the number of communities, l the length of the sides of the considered square 2-dimensional space, μ the graph's density, λ_d the mixing coefficient, $f(d)$ the deterrence function and I_{min}, I_{max} the minimum and maximum intrinsic strengths. We generate graphs according to the following procedure:

1. Attribute a position to each of the N nodes in space, defined uniformly at random such that $n_x \in [0, l]$, $n_y \in [0, l]$
2. Attribute an *intrinsic strength* to each node, uniformly at random such that $n_I \in [I_{min}, I_{max}]$
3. Attribute a community to each node, taken uniformly at random in the set $\{1, \ldots, C\}$
4. Compute p_{ij}^{Inc} for all i,j, for the chosen λ_d,P_{ij}^{SNM},$f(d)$
5. Distribute uniformly at random $L = \mu N(N-1)/2$ edges, where there is an edge between i and j with probability p_{ij}^{Inc}, and multiple edges are interpreted as weights.

2.3 Community Detection Algorithm

The community detection procedure we use is identical to the one in [3].

2.4 Community Partition Evaluation

For each set of benchmark's parameters to test, a graph is generated, and communities are found for each tested null-model using modified Louvain. It then becomes possible to compare the detected partition, result of the algorithm, with the planted partition. As in previous works [3, 7], we use the Normalized Mutual Information (NMI) [11].

3 Definition of a Degree Constrained Gravity-Based Model

In the previously introduced spatial null models, there is no simple relation between the *intrinsic strength* of a node and its actual strength (sum of weights of adjacent edges) in a network generated according to this null model. This means that if the only available data is an observed network, and we use observed degrees of nodes as a proxy for their intrinsic importance, then any of the previously proposed spatial null model fitted on this observed network will not conserve the degrees of nodes. The null model we propose is searching for a degree constrained solution, i.e. a spatial null-model preserving the degrees of nodes.

To do so, we take inspiration from the doubly constrained gravity model [12], and adapt it to the case of spatial networks with estimated deterrence function. The intuition is that we are searching for values of intrinsic strength that would best explain the observed degrees. We present the method in its more general form, adapted to oriented weighted networks. Therefore, we compute separately for each node an *Incoming estimated Intrinsic Strength* (n^{Ieis}) and an *Outgoing estimated Intrinsic Strength* (n^{Oeis}). For non-oriented networks, $n^{Ieis} = n^{Oeis}$.

The method consists in iteratively estimating the new values for n^{Ieis} and n^{Oeis} that satisfies the observed indegrees (deg^{in}) and outdegrees (deg^{out}) constraints.

We can define them recursively as:

$$n^{Ieis} = \frac{deg^{out}(i)}{\sum_i n^{Oeis} f(d_{ij})}, n^{Oeis} = \frac{deg^{in}(i)}{\sum_i n^{Ieis} f(d_{ij})} \tag{6}$$

and the corresponding Degree Constrained gravity model is:

$$P_{ij}^{DCgrav} = n^{Oeis} n^{Ieis} f(d_{ij}) \tag{7}$$

Starting with initial values $n^{Oeis} = deg^{out}$ and $n^{Ieis} = deg^{in}$, we first compute all values for n^{Oeis}, then all values for n^{Ieis}, and so on and so forth until the degrees obtained in the gravity model defined in Eq. 7 are close enough to the target network. Although this process is known to converge [12], in this article we will use a fix number of iterations, $i = 5$, to avoid discussions on stopping criterium and convergence time.

3.1 Recomputation of the Deterrence Function

Because the computed deterrence function depends on the *intrinsic strength* of nodes, estimating it using observed degrees as a proxy leads to a biased approximation. By recomputing the deterrence function after each iteration of the algorithm, we can in part correct this bias.

4 Validation of Null Models on Synthetic Benchmarks

4.1 Benchmark Parameters

To limit the number of cases to study, we decided to fix some parameters. The influence of these parameters has already been studied in [7], and minor changes do not affect much the results. Of course, major changes can have strong effect, for instance is the graph becomes extremely sparse, finding communities becomes harder for all methods.

We choose values close to the ones studied in [7]. Fixed parameters and their values: $N = 100$, $l = 10$, $\mu = 100$, $I_{min} = 10$, $I_{max} = 100$, $C = 2$.

For the deterrence function, whose impact was not studied in [7], we consider several values: For gravity based benchmarks, we take $f(d)$ among $\{f(x) = 1/x, f(x) = 1/x^{0.5}, f(x) = 1/x^2\}$. For the Radiation case, we consider $f(d) \in \{f(x) = 1, f(x) = 1/x\}$. $f(x) = 1$ corresponds to the original definition of the Radiation model, with no explicit definition of deterrence function.

As in [7], we allow the mixing parameter λ_d to vary from 0 to 1, i.e. from perfectly unambiguous community structure to a network with only a spatial structure.

4.2 Evaluation Process

For each set of parameters, we generate 50 instances of networks. For each instance, we run the modified Louvain algorithm with each of the following null models:

- Configuration model [13], noted as NG
- Gravity-based [3], noted as Gra
- Radiation-based (original) [10], noted as Rad
- Radiation-based with deterrence function [7], noted as $DFrad$
- Degree constrained Gravity-based, introduced in the present paper, noted as $DCgra$

For each of these algorithms, we consider two cases, Fully Informed and Network/Position Only.

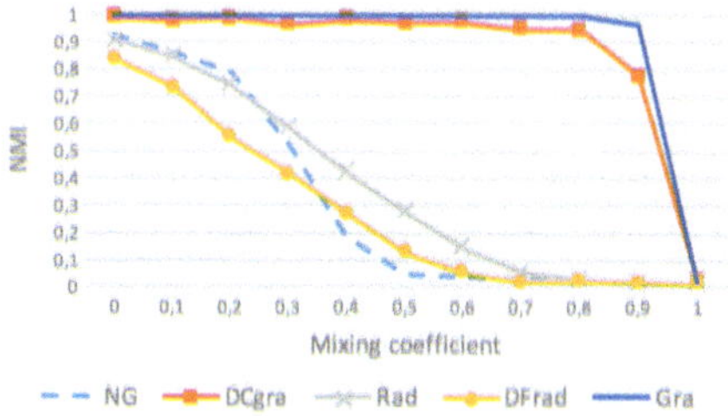

(a) $f(x) = 1/x$, Fully Informed

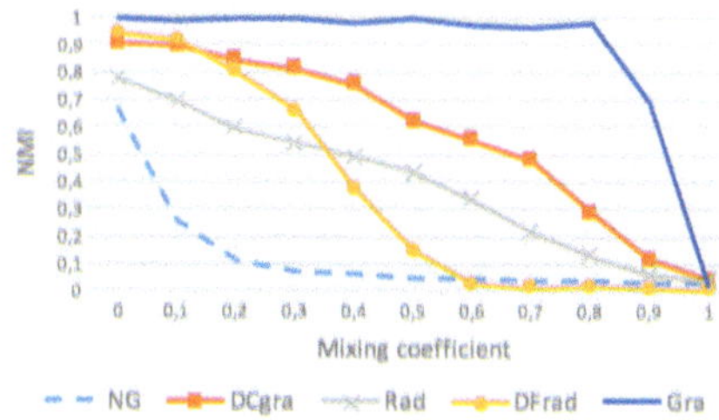

(b) $f(x) = 1/x^2$, Fully Informed

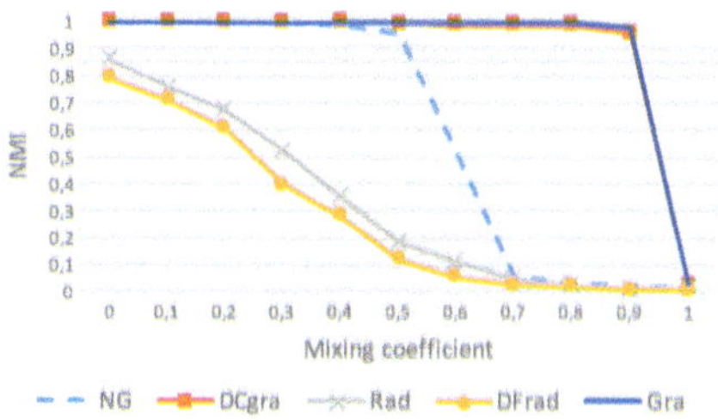

(c) $f(x) = 1/x^{0.5}$, Fully Informed

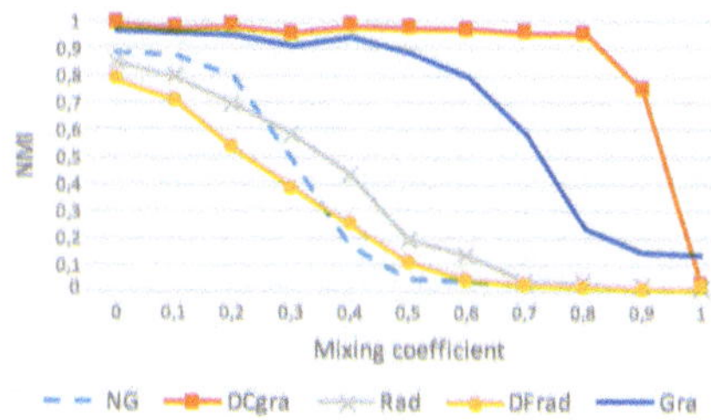

(d) $f(x) = 1/x$, N/P Only

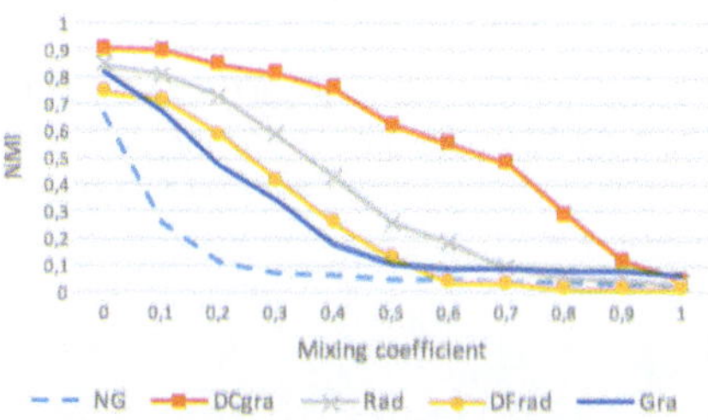

(e) $f(x) = 1/x^2$, N/P Only

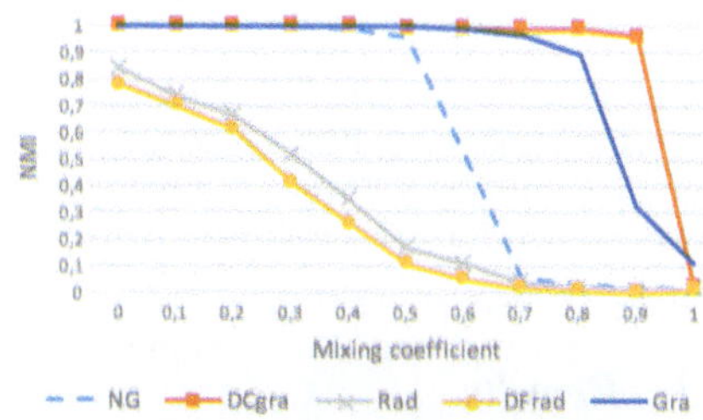

(f) $f(x) = 1/x^{0.5}$, N/P Only

Fig. 1 Results for the synthetic benchmark, using a **generative Gravity model**. In fully informed cases, the gravity null-model is the most efficient, while the proposed DCgra model gives best results when only the network and position of nodes is known

In the **Fully Informed** version, we consider that we know not only the observed network, but also the intrinsic strength of nodes and the deterrence function used to generate the network. This is the same setting than tests conduced in [3, 7].

In the **Network/Position Only** version, we consider that we only know the observed network, and the position of nodes. The deterrence function is first computed from these data, when needed, and the degree of nodes is used as proxy for the intrinsic importance of nodes, as it is often done in applications to collected datasets, for instance in [6, 7]. This setting is more realistic, for applications to real world datasets.

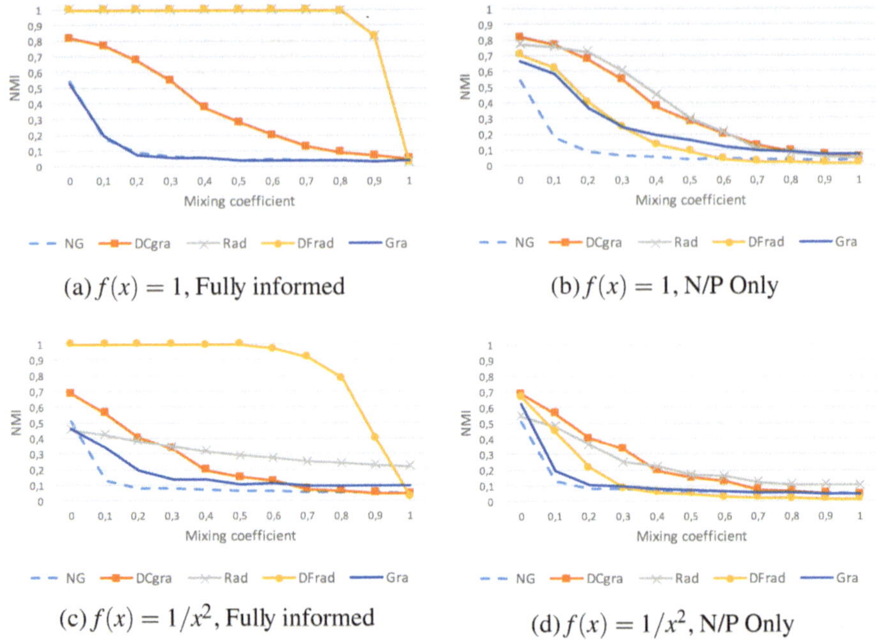

Fig. 2 Results on the synthetic benchmark, using a **Radiation generative model**, both for **Fully Informed** and **Network/Position Only** settings. While the DFrad model gives by far the best results in fully informed cases, its efficacy dwindle when less information is available, and the DCgra model and the original Radiation Null-models give the best results

4.3 Results

In Fig. 1, left column, we present the results for the synthetic benchmark with a generative gravity model, and the fully informed case. As expected, the *Gra* null-model is the most efficient. We can observe that the problem becomes harder with the increase in the exponent of the deterrence function. In fact, the more this exponent is low, the more the network resemble a non-spatial network. The proposed DCgravity model, that does not benefit from full information, comes nevertheless second in most settings.

In Fig. 1, right column, tests are conduced with same settings but in Network/Position Only version, i.e. similar to a collected dataset. In this configuration, results for the original gravity model dwindle, in particular with a high exponent for the deterrence function, in which cases the radiation models give better results. The DCgravity algorithm gives best results in most settings.

In Fig. 2, a radiation generative model is used. With the function $f(x) = 1$, both Rad and DFrad give similar result, because this function is implicitly assumed by the Rad null model. Although they reach high NMI scores in Full Information settings, again the results shrink in the N/P only case, in particular for DFrad. With a modified deterrence function, DFrad is the only one to give good results on Fully informed settings, but again, it does not maintain this efficiency for N/P Only. Interestingly, results for DCgra and Rad are comparable in the N/P Only cases.

Acknowledgements This work is funded in part by the ANR Vel'Innov: Vel'Innov ANR-12-SOIN-0001-02, European Commission H2020 FETPROACT 2016–2017 program under grant 732942 (ODYCCEUS), by the ANR (French National Agency of Research) under grants ANR-15-CE38-0001 (AlgoDiv) and ANR-13-CORD-0017-01 (CODDDE), by the French program "PIA - Usages, services et contenus innovants" under grant O18062-44430 (REQUEST), and by the Ile-de-France program FUI21 under grant 16010629 (iTRAC).

References

1. Denise, A., Vasconcellos, M., Welsh, D.J.: The random planar graph. Congressus Numerantium, 61–80 (1996)
2. Barthélemy, M.: Spatial networks. Phys. Rep. **499**(1), 1–101 (2011)
3. Expert, P., Evans, T., Blondel, V., Lambiotte, R.: Uncovering space-independent communities in spatial networks. Proc. Natl. Acad. Sci. **108**(19), 7663–7668 (2011)
4. Blondel, V., Guillaume, J., Lambiotte, R., Lefebvre, E.: Fast unfolding of communities in large networks. J. Stat. Mech. Theory Exp. **2008**(10), P10008 (2008)
5. Liu, Y., Sui, Z., Kang, C., Gao, Y.: Uncovering patterns of inter-urban trip and spatial interaction from social media check-in data. PloS One **9**(1), e86026 (2014)
6. Austwick, M.Z., OBrien, O., Strano, E., Viana, M.: The structure of spatial networks and communities in bicycle sharing systems. PloS One **8**(9), e74685 (2013)
7. Sarzynska, M., Leicht, E.A., Chowell, G., Porter, M.A.: Null models for community detection in spatially embedded, temporal networks. J. Complex Netw. cnv027 (2015)
8. Newman, M.E.: Modularity and community structure in networks. Proc. Natl. Acad. Sci. **103**(23), 8577–8582 (2006)
9. Lenormand, M., Bassolas, A., Ramasco, J.J.: Systematic comparison of trip distribution laws and models. J. Transp. Geogr. **51**, 158–169 (2016)
10. Simini, F., González, M.C., Maritan, A., Barabási, A.-L.: A universal model for mobility and migration patterns. Nature **484**(7392), 96–100 (2012)
11. Strehl, A., Ghosh, J.: Cluster ensembles—a knowledge reuse framework for combining multiple partitions. J. Mach. Learn. Res. **3**, 583–617, Dec. 2002
12. Williams, I.: A comparison of some calibration techniques for doubly constrained models with an exponential cost function. Transp. Res. **10**(2), 91–104 (1976)
13. Newman, M.E., Strogatz, S.H., Watts, D.J.: Random graphs with arbitrary degree distributions and their applications. Phys. Rev. E **64**(2), 026118 (2001)

deterrence function, D(d) is the only one to give good results. [illegible] ... that modularity [illegible] for NPP [illegible] ... [illegible] ... only [illegible]

Node-Centric Community Detection in Multilayer Networks with Layer-Coverage Diversification Bias

R. Interdonato, A. Tagarelli, D. Ienco, A. Sallaberry and P. Poncelet

Abstract The problem of node-centric, or *local*, community detection in information networks refers to the identification of a community for a given input node, having limited information about the network topology. Existing methods for solving this problem, however, are not conceived to work on complex networks. In this paper, we propose a novel framework for local community detection based on the multilayer network model. Our approach relies on the maximization of the ratio between the community internal connection density and the external connection density, according to multilayer similarity-based community relations. We also define a biasing scheme that allows the discovery of local communities characterized by different degrees of layer-coverage diversification. Experimental evaluation conducted on real-world multilayer networks has shown the significance of our approach.

1 Introduction

The classic problem of community detection in a network graph corresponds to an optimization problem which is *global* as it requires knowledge on the *whole* network structure. The problem is known to be computationally difficult to solve, while its

---

R. Interdonato (✉) · A. Tagarelli (✉)
DIMES - University of Calabria, Rende, Italy
e-mail: andrea.tagarelli@dimes.unical.it

R. Interdonato
e-mail: rinterdonato@dimes.unical.it

D. Ienco
IRSTEA - UMR TETIS, Montpellier, France
e-mail: dino.ienco@irstea.fr

A. Sallaberry
LIRMM - Université Paul Valéry, Montpellier, France
e-mail: arnaud.sallaberry@lirmm.fr

P. Poncelet
LIRMM - Université de Montpellier, Montpellier, France
e-mail: pascal.poncelet@lirmm.fr

B. Gonçalves et al. (eds.), *Complex Networks VIII*,
Springer Proceedings in Complexity, DOI 10.1007/978-3-319-54241-6_5

approximate solutions have to cope with both accuracy and efficiency issues that become more severe as the network increases in size. Large-scale, web-based environments have indeed traditionally represented a natural scenario for the development and testing of effective community detection approaches. In the last few years, the problem has attracted increasing attention in research contexts related to *complex networks* [2, 7–9, 11–14], whose modeling and analysis is widely recognized as a useful tool to better understand the characteristics and dynamics of multiple, interconnected types of node relations and interactions [1, 6].

Nevertheless, especially in social computing, one important aspect to consider is that we might often want to identify the personalized network of social contacts of interest to a single user only. To this aim, we would like to determine the expanded neighborhood of that user which forms a densely connected, relatively small subgraph. This is known as *local community detection* problem [4, 5], whose general objective is, given limited information about the network, to identify a community structure which is centered on one or few seed users. Existing studies on this problem have focused, however, on social networks that are built on a single user relation type or context [4, 15]. As a consequence, they are not able to profitably exploit the fact that most individuals nowadays have multiple accounts across different social networks, or that relations of different types (i.e., online as well as offline relations) can be available for the same population of a social network [6].

In this work, we propose a novel framework based on the multilayer network model for the problem of local community detection, which overcomes the aforementioned limitations in the literature, i.e., community detection on a multilayer network but from a global perspective, and local community detection but limited to monoplex networks. We have recently brought the local community detection problem into the context of multilayer networks [10], by providing a preliminary formulation based on an unsupervised approach. A key aspect of our proposal is the definition of similarity-based community relations that exploit both internal and external connectivity of the nodes in the community being constructed for a given seed, while accounting for different layer-specific topological information. Here we push forward our research by introducing a parametric control in the similarity-based community relations for the layer-coverage diversification in the local community being discovered. Our experimental evaluation conducted on three real-world multilayer networks has shown the significance of our approach.

2 Multilayer Local Community Detection

2.1 *The* **ML-LCD** *Method*

We refer to the multilayer network model described in [9]. We are given a set of layers $\mathscr{L}$ and a set of entities (e.g., users) $\mathscr{V}$. We denote with $G_{\mathscr{L}} = (V_{\mathscr{L}}, E_{\mathscr{L}}, \mathscr{V}, \mathscr{L})$ the multilayer graph such that $V_{\mathscr{L}}$ is a set of pairs $v \in \mathscr{V}$, $L \in \mathscr{L}$, and $E_{\mathscr{L}} \subseteq V_{\mathscr{L}} \times V_{\mathscr{L}}$

is the set of undirected edges. Each entity of V appears in at least one layer, but not necessarily in all layers. Moreover, in the following we will consider the specific case for which nodes connected through different layers the same entity in $\mathscr{V}$, i.e., $G_{\mathscr{L}}$ is a multiplex graph.

Local community detection approaches generally implement some strategy that at each step considers a node from one of three sets, namely: the community under construction (initialized with the seed node), the "shell" of nodes that are neighbors of nodes in the community but do not belong to the community, and the unexplored portion of the network. A key aspect is hence how to select the *best* node in the shell to add to the community to be identified. Most algorithms, which are designed to deal with monoplex graphs, try to maximize a function in terms of the *internal* edges, i.e., edges that involve nodes in the community, and to minimize a function in terms of the *external* edges, i.e., edges to nodes outside the community. By accounting for both types of edges, nodes that are candidates to be added to the community being constructed are penalized in proportion to the amount of links to nodes external to the community [5]. Moreover, as first analyzed in [4], considering the internal-to-external *connection density* ratio (rather than the absolute amount of internal and external links to the community) allows for alleviating the issue of inserting many weakly-linked nodes (i.e., *outliers*) into the local community being discovered. In this work we follow the above general approach and extend it to identify local communities over a multilayer network.

Given $G_{\mathscr{L}} = (V_{\mathscr{L}}, E_{\mathscr{L}}, \mathscr{V}, \mathscr{L})$ and a seed node v_0, we denote with $C \subseteq \mathscr{V}$ the node set corresponding to the local community being discovered around node v_0; moreover, when the context is clear, we might also use C to refer to the local community subgraph. We denote with $S = \{v \in \mathscr{V} \setminus C \mid \exists((u, L_i), (v, L_j)) \in E_{\mathscr{L}} \wedge u \in C\}$ the *shell* set of nodes outside C, and with $B = \{u \in C \mid \exists((u, L_i), (v, L_j)) \in E_{\mathscr{L}} \wedge v \in S\}$ the *boundary* set of nodes in C.

Our proposed method, named **M**ulti**L**ayer **L**ocal **C**ommunity **D**etection (ML-LCD), takes as input the multilayer graph $G_{\mathscr{L}}$ and a seed node v_0, and computes the local community C associated to v_0 by performing an iterative search that seeks to maximize the value of *similarity-based local community function* for C ($LC(C)$), which is obtained as the ratio of an *internal community relation* $LC^{int}(C)$ to an *external community relation* $LC^{ext}(C)$. We shall formally define these later in Sect. 2.2.

Algorithm ML-LCD works as follows. Initially, the boundary set B and the community C are initialized with the starting seed, while the shell set S is initialized with the neighborhood set of v_0 considering all the layers in $\mathscr{L}$. Afterwards, the algorithm computes the initial value of $LC(C)$ and starts expanding the node set in C: it evaluates all the nodes v belonging to the current shell set S, then selects the vertex v^* that maximizes the value of $LC(C)$. The algorithm checks if *(i)* v^* actually increases the quality of C (i.e., $LC(C \cup \{v^*\}) > LC(C)$) and *(ii)* v^* helps to strength the internal connectivity of the community (i.e., $LC^{int}(C \cup \{v^*\}) > LC^{int}(C)$). If both conditions are satisfied, node v^* is added to C and the shell set is updated accordingly, otherwise node v^* is removed from S as it cannot lead to an increase in the value of $LC(C)$. In any case, the boundary set B and $LC(C)$ are updated. The algorithm terminates when no further improvement in $LC(C)$ is possible.

2.2 Similarity-Based Local Community Function

To account for the multiplicity of layers, we define the multilayer local community function $LC(\cdot)$ based on a notion of similarity between nodes. In this regard, two major issues are how to choose the analytical form of the similarity function, and how to deal with the different, layer-specific connections that any two nodes might have in the multilayer graph. We address the first issue in an unsupervised fashion, by resorting to any similarity measure that can express the topological affinity of two nodes in a graph. Concerning the second issue, one straightforward solution is to determine the similarity between any two nodes focusing on each layer at a time. The above points are formally captured by the following definitions. We denote with E^C the set of edges between nodes that belong to C and with E_i^C the subset of E^C corresponding to edges in a given layer L_i. Analogously, E^B refers to the set of edges between nodes in B and nodes in S, and E_i^B to its subset corresponding to L_i.

Given a community C, we define the *similarity-based local community function* $LC(C)$ as the ratio between the *internal community relation* and *external community relation*, respectively defined as:

$$LC^{int}(C) = \frac{1}{|C|} \sum_{v \in C} \sum_{L_i \in \mathscr{L}} \sum_{(u,v) \in E_i^C \ \wedge \ u \in C} sim_i(u, v) \tag{1}$$

$$LC^{ext}(C) = \frac{1}{|B|} \sum_{v \in B} \sum_{L_i \in \mathscr{L}} \sum_{(u,v) \in E_i^B \ \wedge \ u \in S} sim_i(u, v) \tag{2}$$

In the above equations, function $sim_i(u, v)$ computes the similarity between any two nodes u, v contextually to layer L_i. In this work, we define it in terms of Jaccard coefficient, i.e., $sim_i(u, v) = \frac{|N_i(u) \cap N_i(v)|}{|N_i(u) \cup N_i(v)|}$, where $N_i(u)$ denotes the set of neighbors of node u in layer L_i.

2.3 Layer-Coverage Diversification Bias

When discovering a multilayer local community centered on a seed node, the iterative search process in ML-LCD that seeks to maximize the similarity-based local community measure, explores the different layers of the network. This implies that the various layers might contribute very differently from each other in terms of edges constituting the local community structure. In many cases, it can be desirable to control the degree of heterogeneity of relations (i.e., layers) inside the local community being discovered.

In this regard, we identify two main approaches:

- **Diversification-oriented approach**. This approach relies on the assumption that a local community is better defined by increasing as much as possible the number of edges belonging to different layers. More specifically, we might want to obtain a local community characterized by high diversification in terms of presence of layers and variability of edges coming from different layers.

- **Balance-oriented approach**. Conversely to the previous case, the aim is to produce a local community that shows a certain *balance* in the presence of layers, i.e., low variability of edges over the different layers. This approach relies on the assumption that a local community might be well suited to real cases when it is uniformly distributed among the different edge types taken into account.

Following the above observations, here we propose a methodology to incorporate a parametric control of the layer-coverage diversification in the local community being discovered. To this purpose, we introduce a *bias factor* β in ML-LCD which impacts on the node similarity measure according to the following logic:

$$\beta = \begin{cases} (0, 1], & \textit{diversification-oriented bias} \\ 0, & \textit{no bias} \\ [-1, 0), & \textit{balance-oriented bias} \end{cases} \tag{3}$$

Positive values of β push the community expansion process towards a diversification-oriented approach, and, conversely, negative β lead to different levels of balance-oriented scheme. Note that the *no bias* case corresponds to handling the node similarity "as is". Note also that, by assuming values in a continuous range, at each iteration ML-LCD is enabled to make a decision by accounting for a wider spectrum of degrees of layer-coverage diversification.

Given a node $v \in B$ and a node $u \in S$, for any $L_i \in \mathcal{L}$, we define the β-biased similarity $sim_{\beta,i}(u, v)$ as follows:

$$sim_{\beta,i}(u, v) = \frac{2sim_i(u, v)}{1 + e^{-bf}}, \tag{4}$$

$$bf = \beta[f(C \cup \{u\}) - f(C)] \tag{5}$$

where bf is a *diversification factor* and $f(C)$ is a function that measures the current diversification between the different layers in the community C; in the following, we assume it is defined as the standard deviation of the number of edges for each layer in the community. The difference $f(C \cup \{u\}) - f(C)$ is positive when the insertion of node u into the community increases the coverage over a subset of layers, thus diversifying the presence of layers in the local community. Consequently, when β is positive, the diversification effect is desired, i.e., there is a boost in the value of $sim_{\beta,i}$ (and vice versa for negative values of β). Note that β introduces a bias on the similarity between two nodes only when evaluating the inclusion of a shell node into a community C, i.e., when calculating $LC^{ext}(C)$.

3 Experimental Evaluation

We used three multilayer network datasets, namely *Airlines* (417 nodes corresponding to airport locations, 3588 edges, 37 layers corresponding to airline companies) [3], *AUCS* (61 employees as nodes, 620 edges, 5 acquaintance relations as layers) [6], and *RealityMining* (88 users as nodes, 355 edges, 3 media types employed to communicate as layers) [8]. All network graphs are undirected, and inter-layer links are regarded as coupling edges.

Size and structural characteristics of local communities. We first analyzed the size of the local communities extracted by ML-LCD for each node. Table 1 reports on the mean and standard deviation of the size of the local communities by varying of β. As regards the *no bias* solution (i.e., $\beta = 0.0$), largest local communities correspond to *Airlines* (mean 11.33 ± 14.78), while medium size communities (7.90 ± 2.74) are found for *AUCS* and relatively small communities (3.37 ± 1.77) for *RealityMining*. The impact of β on the community size is roughly proportional to the number of layers, i.e., high on *Airlines*, medium on *AUCS* and low on *RealityMining*. For *Airlines* and *AUCS*, smallest communities are obtained with the solution corresponding to $\beta = -1.0$, thus suggesting that the discovery process becomes more xenophobic (i.e., less inclusive) while shifting towards a balance-oriented scheme. Moreover, on *Airlines*, the mean size follows a roughly normal distribution, with most inclusive solution (i.e., largest size) corresponding to the unbiased one. A near normal distribution (centered on $0.2 \leq \beta \leq 0.4$) is also observed for *RealityMining*, while mean size values linearly increase with β for *AUCS*.

To understand the effect of β on the structure of the local communities, we analyzed the distributions of per-layer mean *average path length* and mean *clustering coefficient* of the identified communities (results not shown). One major remark is that on the networks with a small number of layers, the two types of distributions tend to follow an increasing trend for balance-oriented bias (i.e., negative β), which becomes roughly constant for the diversification-oriented bias (i.e., positive β). On *Airlines*, variability happens to be much higher for some layers, which in the case of mean average path length ranges between 0.1 and 0.5 (as shown by a rapidly decreasing trend for negative β, followed by a peak for $\beta = 0.2$, then again a decreasing trend).

Distribution of layers over communities. We also studied how the bias factor impacts on the distribution of number of layers over communities, as shown in Fig. 1. This analysis confirmed that using positive values of β produces local communities that lay on a higher number of layers. This outcome can be easily explained since positive values of β favor the inclusion of nodes into the community which increase layer-coverage diversification, thus enabling the exploration of further layers also in an advanced phase of the discovering process. Conversely, negative values of β are supposed to yield a roughly uniform distribution of the layers which are covered by the community, thus preventing the discovery process from including nodes coming from unexplored layers once the local community is already characterized by a certain subset of layers.

Table 1 Mean and standard deviation size of communities by varying β (with step of 0.1)

Dataset		−1.0	−0.9	−0.8	−0.7	−0.6	−0.5	−0.4	−0.3	−0.2	−0.1	*0.0*	0.1	0.2	0.3	0.4	0.5	0.6	0.7	0.8	0.9	1.0
Airlines	Mean	5.73	5.91	6.20	6.47	6.74	7.06	7.57	8.10	9.13	10.33	*11.33*	9.80	9.02	8.82	8.37	8.20	7.93	7.53	7.26	7.06	7.06
	sd	4.68	4.97	5.45	5.83	6.39	6.81	7.63	8.62	10.58	12.80	*14.78*	12.10	10.61	10.07	9.39	9.15	8.67	7.82	7.46	7.35	7.27
AUCS	Mean	6.38	6.59	6.64	6.75	6.84	6.85	6.92	7.13	7.16	7.77	*7.90*	8.77	8.92	8.92	8.89	8.89	8.89	8.87	8.85	8.85	8.85
	sd	1.48	1.51	1.59	1.69	1.85	1.85	1.87	2.15	2.18	2.40	*2.74*	3.16	3.33	3.33	3.27	3.27	3.27	3.26	3.23	3.23	3.23
Reality-mining	Mean	3.21	3.24	3.25	3.25	3.32	3.32	3.34	3.34	3.34	3.37	*3.37*	3.38	3.39	3.39	3.39	3.36	3.36	3.32	3.18	3.17	3.17
	sd	1.61	1.64	1.66	1.66	1.73	1.73	1.74	1.74	1.74	1.77	*1.77*	1.78	1.78	1.78	1.78	1.74	1.74	1.71	1.60	1.59	1.59

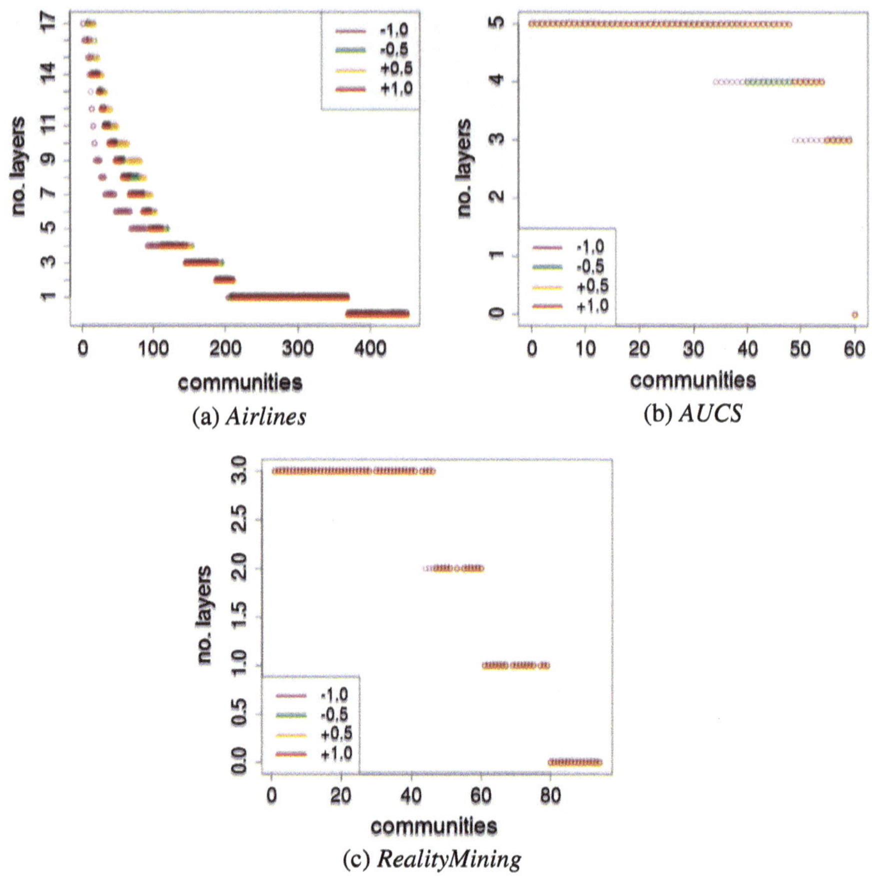

Fig. 1 Distribution of number of layers over communities by varying β. Communities are sorted by decreasing number of layers

As regards the effects of the bias factor on the layer-coverage diversification, we analyzed the standard deviation of the per-layer number of edges by varying β (results not shown, due to space limits of this paper). As expected, standard deviation values are roughly proportional to the setting of the bias factor for all datasets. Considering the local communities obtained with negative β, the layers on which they lay are characterized by a similar presence (in terms of number of edges) in the induced community subgraph. Conversely, for the local communities obtained using positive β, the induced community subgraph may be characterized by a small subset of layers, while other layers may be present with a smaller number of relations.

Similarity between communities. The smooth effect due to the diversification-oriented bias is confirmed when analyzing the similarity between the discovered local communities. Figure 2 shows the average Jaccard similarity between solutions obtained by varying β (i.e., in terms of nodes included in each local community). Jac-

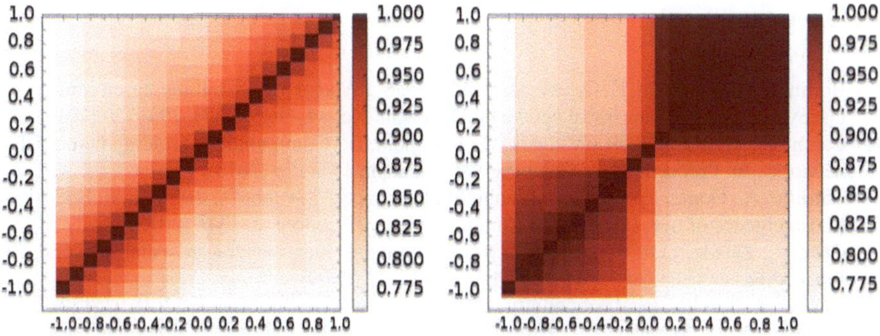

(a) *Airlines* (b) *AUCS*

Fig. 2 Average Jaccard similarity between solutions obtained by varying β

card similarities vary in the range [0.75, 1.0] for *AUCS* and *Airlines*, and in the range [0.9, 1.0] for *RealityMining* (results not shown). For datasets with a lower number of layers (i.e., *AUCS* and *RealityMining*), there is a strong separation between the solutions obtained for $\beta > 0$ and the ones obtained with $\beta < 0$. On *AUCS*, the local communities obtained using a diversification-oriented bias show Jaccard similarities close to 1, while there is more variability among the solutions obtained with the balance-oriented bias. Effects of the bias factor are lower on *RealityMining*, with generally high Jaccard similarities. On *Airlines*, the effects of the bias factor are still present but smoother, with gradual similarity variations in the range [0.75, 1.0].

4 Conclusion

We addressed the novel problem of local community detection in multilayer networks, providing a greedy heuristic that iteratively attempts to maximize the internal-to-external connection density ratio by accounting for layer-specific topological information. Our method is also able to control the layer-coverage diversification in the local community being discovered, by means of a bias factor embedded in the similarity-based local community function. Evaluation was conducted on real-world multilayer networks. As future work, we plan to study alternative objective functions for the ML-LCD problem. It would also be interesting to enrich the evaluation part based on data with ground-truth information. We also envisage a number of application problems for which ML-LCD methods can profitably be used, such as friendship prediction, targeted influence propagation, and more in general, mining in incomplete networks.

References

1. Berlingerio, M., Pinelli, F., Calabrese, F.: ABACUS: frequent pattern mining-based community discovery in multidimensional networks. Data Min. Knowl. Disc. **27**(3), 294–320 (2013)
2. Carchiolo, V., Longheu, A., Malgeri, M., Mangioni, G.: Communities unfolding in multislice networks. In: Proceedings of Complex Networks, pp. 187–195 (2010)
3. Cardillo, A., Gomez-Gardenes, J., Zanin, M., Romance, M., Papo, D., del Pozo, F., Boccaletti, S.: Emergence of network features from multiplexity. Sci. Rep. **3**, 1344 (2013)
4. Chen, J., Zaïane, O.R., Goebel, R.: Local community identification in social networks. In: Proceedings of IEEE/ACM ASONAM, pp. 237–242 (2009)
5. Clauset, A.: Finding local community structure in networks. Phys. Rev. E **72**(2), 026132 (2005)
6. Dickison, M.E., Magnani, M., Rossi, L.: Multilayer Social Networks. Cambridge University Press (2016)
7. De Domenico, M., Lancichinetti, A., Arenas, A., Rosvall, M.: Identifying modular flows on multilayer networks reveals highly overlapping organization in social systems. Phys. Rev. X **5**(1), 011027 (2015)
8. Kim, J., Lee, J.-G.: Community detection in multi-layer graphs: a survey. SIGMOD Record **44**(3), 37–48 (2015)
9. Kivela, M., Arenas, A., Barthelemy, M., Gleeson, J.P., Moreno, Y., Porter, M.A.: Mutilayer networks. J. Complex Netw. **2**(3), 203–271 (2014)
10. Interdonato, R., Tagarelli, A., Ienco, D., Sallaberry, A., Poncelet, P.: Local community detection in multilayer networks. In: Proceedings of IEEE/ACM ASONAM, pp. 1382–1383 (2016)
11. Loe, C.W., Jensen, H.J.: Comparison of communities detection algorithms for multiplex. Phys. A **431**, 29–45 (2015)
12. Mucha, P.J., Richardson, T., Macon, K., Porter, M.A., Onnela, J.-P.: Community structure in time-dependent, multiscale, and multiplex networks. Science **328**(5980), 876–878 (2010)
13. Papalexakis, E.E., Akoglu, L., Ienco, D.: Do more views of a graph help? Community detection and clustering in multi-graphs. In: Proceedings of Fusion, pp. 899–905 (2013)
14. Peixoto, T.P.: Inferring the mesoscale structure of layered, edge-valued, and time-varying networks. Phys. Rev. E **92**(4), 042807 (2015)
15. Zakrzewska, A., Bader, D.A.: A dynamic algorithm for local community detection in graphs. In: Proceedings of IEEE/ACM ASONAM, pp. 559–564 (2015)

Community Detection in Signed Networks Based on Extended Signed Modularity

Tsuyoshi Murata, Takahiko Sugihara and Talel Abdessalem

Abstract Community detection is important for analyzing and visualizing given networks. In real world, many complex systems can be modeled as signed networks composed of positive and negative edges. Although community detection in signed networks has been attempted by many researchers, studies for detecting detailed structures remain to be done. In this paper, we extend modularity for signed networks, and propose a method for optimizing our modularity, which is an efficient hierarchical agglomeration algorithm for detecting communities in signed networks. Based on the experiments with large-scale real world signed networks such as Wikipedia, Slashdot and Epinions, our method enables us to detect communities and inner factions inside the communities.

Keywords Signed networks · Community detection · Signed modularity

1 Introduction

Communities in networks are defined as the groups of nodes within which the edges are dense but between which the edges are sparse. Community detection in networks attracts many researchers, and many methods for community detection are proposed [1, 4, 9]. Most of the previous research on community detection are for normal networks composed of only one edge type. However, several real relations can be represented as signed networks composed of positive and negative edges. In this paper, we extend the modularity for signed networks in order to detect communities

T. Murata (✉) · T. Sugihara
Department of Computer Science, School of Computing,
Tokyo Institute of Technology, W8-59 2-12-1 Ookayama, Meguro, Tokyo 152-8552, Japan
e-mail: murata@c.titech.ac.jp

T. Abdessalem
Computer Science and Networks Department, Telecom ParisTech,
46 rue Barrault, 75013 Paris, France
e-mail: talel.abdessalem@telecom-paristech.fr

B. Gonçalves et al. (eds.), *Complex Networks VIII*,
Springer Proceedings in Complexity, DOI 10.1007/978-3-319-54241-6_6

in signed networks. It is composed of positive and negative modularity, and it includes a balancing parameter for the importance of both types of edges.

Moreover, we propose a method for optimizing our modularity, which is an efficient hierarchical agglomeration algorithm for detecting communities in signed networks. It is based on an efficient optimization method for normal networks proposed by Clauset et al. [1].

We apply our method to several signed networks which represent the relationships among users on websites such as Wikipedia, Slashdot and Epinions. We successfully detect communities in large-scale signed networks which have more than 60,000 nodes and more than 600,000 edges. Our method can control the result by adjusting a parameter and it enables us to detect communities and inner factions inside the communities.

2 Related Works

Social relations with friendship and hostility can be represented as signed networks with positive and negative edges. Many attempts have been made for analyzing signed networks. Although structural balance of triangles of positive and negative edges is one of the important topics in signed networks, we will not discuss in this paper. Gomez et al. [5] extended Newman modularity for the analysis of directed and signed networks. Although the proposed modularity is similar to ours, it has the weakness that the balancing factor of positive and negative edges is fixed. Szell et al. [11] analyzes interactions of massive multiplayer online games. The authors claim that reciprocity and clustering coefficient of positive edges are quite different from those of negative edges. They propose STC model for generating triangles from wedges, and the model fit well for the data of online games. Leskovec et al. [6] focus on a task of edge sign prediction, and they propose a method for predicting positive and negative edges based on logistic regression classifier. Maniu et al. [8] built signed network from the interactions of editors in Wikipedia. The dataset called Wikisigned is available at http://konect.uni-koblenz.de/networks/wikisigned-k2. Esmailian et al. [3] discuss the method for detecting communities based on extended Potts Model. Their approach is flow-based, and it is quite different from our modularity optimization-based approach. Influence maximization in signed networks is studied by Li et al. [7], and link recommendation algorithm is proposed by Song et al. [10], which are not focus of this paper.

3 Extended Modularity for Signed Networks

In good partitions of signed networks, positive edges should be dense within communities and sparse between communities, and negative edges should be sparse between communities and dense within communities. We define an extended modularity for undirected signed networks Q_{signed} as a linear combination of positive and negative modularity.

$$Q_{signed} = \alpha Q^{+} - (1-\alpha)Q^{-} \tag{1}$$

Q^{+} in Eq. (1) is a positive modularity, which represents the fraction of the positive edges that fall within the given groups minus the expected such fraction if positive edges were distributed at random. This is represented by Eq. (2).

$$Q^{+} = \frac{1}{2m^{+}} \sum_{ij} (A_{ij}^{+} - \frac{k_i^{+} k_j^{+}}{2m^{+}}) \delta(c_i, c_j) \tag{2}$$

In Eq. (2), m^{+} is the number of positive edges, A^{+} is a positive adjacency matrix and A_{ij}^{+} is its (i, j)-th element.

$$A_{ij}^{+} = \begin{cases} 1 & \text{if there is an positive edge between node } i \text{ and } j \\ 0 & \text{otherwise} \end{cases} \tag{3}$$

The positive degree k_i^{+} of node i is defined as the number of positive edges that connect to i.

$$k_i^{+} = \sum_{j} A_{ij}^{+} \tag{4}$$

Q^{+} in Eq. (1) is negative modularity, which is represented by Eq. (5).

$$Q^{-} = \frac{1}{2m^{-}} \sum_{ij} (A_{ij}^{-} - \frac{k_i^{-} k_j^{-}}{2m^{-}}) \delta(c_i, c_j) \tag{5}$$

In Eq. (5), m^{-} is the number of negative edges, A^{-} is a negative adjacency matrix and A_{ij}^{-} is its (i, j)-th element.

$$A_{ij}^{-} = \begin{cases} 1 & \text{there is an negative edge between node } i \text{ and } j \\ 0 & \text{otherwise} \end{cases} \tag{6}$$

The negative degree k_i^{-} of node i is defined as the number of negative edges that connect to i.

$$k_i^{-} = \sum_{j} A_{ij}^{-} \tag{7}$$

To simplify the description of our algorithm, we define the following four quantities.

$$e_{ij}^{+} = \frac{1}{2m^{+}} \sum_{s \in c_i} \sum_{t \in c_j} A_{st}^{+} \tag{8}$$

$$e_{ij}^{-} = \frac{1}{2m^{-}} \sum_{s \in c_i} \sum_{t \in c_j} A_{st}^{-} \tag{9}$$

$$a_i^{+} = \frac{1}{2m^{+}} \sum_{s \in c_i, t \in V} A_{st}^{+} \tag{10}$$

$$a_i^{-} = \frac{1}{2m^{-}} \sum_{s \in c_i, t \in V} A_{st}^{-} \tag{11}$$

Equation (8) is the fraction of positive edges that connect nodes in community i and nodes in community j, and Eq. (9) is the fraction of negative edges that connect nodes in community i and nodes in community j. Equation (10) is the fraction of positive edges that are attached to nodes in community i, and Eq. (11) is the fraction of negative edges that are attached to nodes in community i.

With the above four quantities, Eqs. (2) and (5) can be simplified as follows:

$$Q^{+} = \sum_i \{e_{ii}^{+} - (a_i^{+})^2\} \tag{12}$$

$$Q^{-} = \sum_i \{e_{ii}^{-} - (a_i^{-})^2\} \tag{13}$$

In good partition of signed networks, the value of Q^{+} should be large and value of Q^{-} should be small. In Eq. (1), α indicates the importance of positive edges and $1 - \alpha$ indicates the importance of negative edges. The value of Q_{signed} is less than 1, and it can be negative value.

Other modularities for signed networks are proposed by Gomez et al. [5] and by Traag et al. [12]. In the modularity by Gomez et al., the value of α is fixed as the proportion of positive edges in the signed network. Parameters are introduced in the modularity by Traag et al. for adjusting the size of communities. They are different from parameter α in our modularity.

For example, the partition of a signed network (Fig. 1) which gives the largest ΔQ_{signed} depends on the value of α. In this signed network, four kinds of partition can be obtained (Fig. 2). When the value of α is large, positive edges within communities

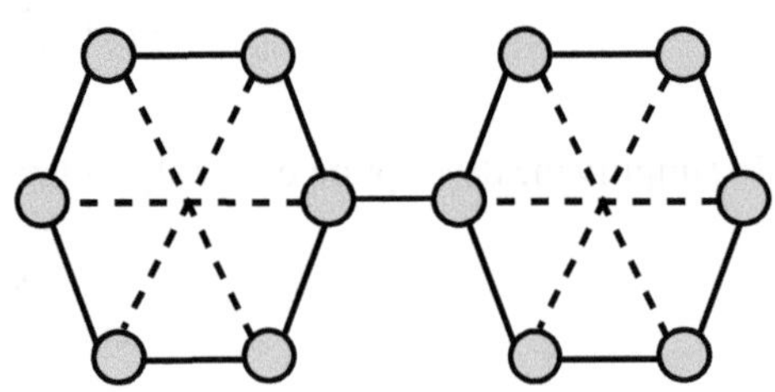

Fig. 1 Example of signed network

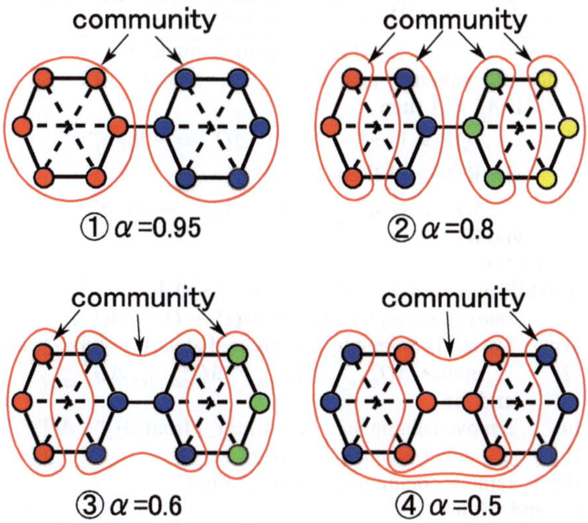

Fig. 2 Communities for different α

are more focused. As a result, partitions that include more negative edges within communities are allowed. Conversely, when the value of α is small, negative edges between communities are more focused. As a result, partitions that include more positive edges between communities are allowed. The modularity by Gomez does not have such flexibility because its parameter is fixed.

4 Method for Optimization

Community detection methods for normal networks cannot be applied directly, because there are negative edges in signed networks. In this paper, we propose a detection method for signed networks based on CNM (the method proposed by Clauset et al.) [1]. By considering the connection patterns of the edges, CNM efficiently calculates the changes in Q that would result from the agglomeration of each pair of communities. CNM cannot be applied directly because there are some connection patterns that do not exist in normal networks. Thus, it is necessary to extend CNM appropriately for signed networks.

In initial state of our method, each node is the sole member of a community. Then, our method calculates changes in Q_{signed} that would result from the agglomeration of each pair of communities connected by positive or negative edges. Until the largest ΔQ_{signed} becomes negative, it continues agglomeration.

Algorithm 1 CNM for signed networks

1: Let $C = \{1, 2, \ldots, n\}$ be the set of communities
2: **for** all community pairs **do** $i = 1, 2, \ldots, n, j = 1, 2, \ldots, n$
3: **if** $A_{ij}^{+} = 1$ **then**
4: $\Delta Q_{ij}^{+} = 1/m^{+} - 2a_i^{+}a_j^{+}, \Delta Q_{ij}^{-} = -2a_i^{-}a_j^{-}$
5: **else if** $A_{ij}^{-} = 1$ **then**
6: $\Delta Q_{ij}^{+} = -2a_i^{+}a_j^{+}, \Delta Q_{ij}^{-} = 1/m^{-} - 2a_i^{-}a_j^{-}$
7: **end if**
8: **end for**
9: **while** max $(\alpha\Delta Q^{+} - (1-\alpha)\Delta Q^{-}) > 0$ **do**
10: $(max\_i, max\_j) = \text{argmax } (\alpha Q^{+} - (1-\alpha)Q^{-})$
11: **for** each community x connected to max_i or max_j **do**
12: update $\Delta Q_{max\_ix}^{+}, \Delta Q_{xmax\_i}^{+}\Delta Q_{max\_ix}^{-}, \Delta Q_{xmax\_i}^{-}$
13: **end for**
14: remove column max_j, row max_j from $\Delta Q^{+}, \Delta Q^{-}$
15: $C \cdot max\_i = max\_i \cup max\_j$
16: remove community $C \cdot max_j$ from C
17: **end while**

In line 4 and 6 of Algorithm 1, ΔQ_{ij}^{+} (an element of ΔQ^{+}) and ΔQ_{ij}^{-} (an element of ΔQ^{-}) are defined as follows:

$$\Delta Q_{ij}^{+} = 2(e_{ij}^{+} - a_i^{+}a_j^{+}) \tag{14}$$

$$\Delta Q_{ij}^{-} = 2(e_{ij}^{-} - a_i^{-}a_j^{-}) \tag{15}$$

If node i and j are connected by a positive edge, then we substitute $e_{ij}^{+} = 1/2m^{+}$, $e_{ij}^{-} = 0$ for Eqs. (14) and (15). If node i and j are connected by a negative edge, then we substitute $e_{ij}^{+} = 0$, $e_{ij}^{-} = 1/2m^{-}$ for Eqs. (14) and (15). Elements of ΔQ^{+} and ΔQ^{-} are as follows:

$$\Delta Q_{ij}^{+} = \begin{cases} 1/m^{+} - 2a_i^{+}a_j^{+} & \text{if } A_{ij}^{+} = 1 \\ -2a_i^{+}a_j^{+} & \text{if } A_{ij}^{-} = 1 \\ 0 & \text{otherwise} \end{cases} \tag{16}$$

$$\Delta Q_{ij}^{-} = \begin{cases} -2a_i^{-}a_j^{-} & \text{if } A_{ij}^{+} = 1 \\ 1/m^{-} - 2a_i^{-}a_j^{-} & \text{if } A_{ij}^{-} = 1 \\ 0 & \text{otherwise} \end{cases} \tag{17}$$

In line 11–14 of Algorithm 1, the community pair (i, j) that gives the largest increase in $\Delta Q_{signed}(= \alpha\Delta Q^{+} - (1-\alpha)\Delta Q^{-})$ is agglomerated. Then, we update column i and row i, and remove column j and row j.

In our method, we avoid the calculation of e_{ij}^{+} and e_{ij}^{-} in Eqs. (14) and (15) because it is time consuming. Considering edge connection pattern between agglomerated

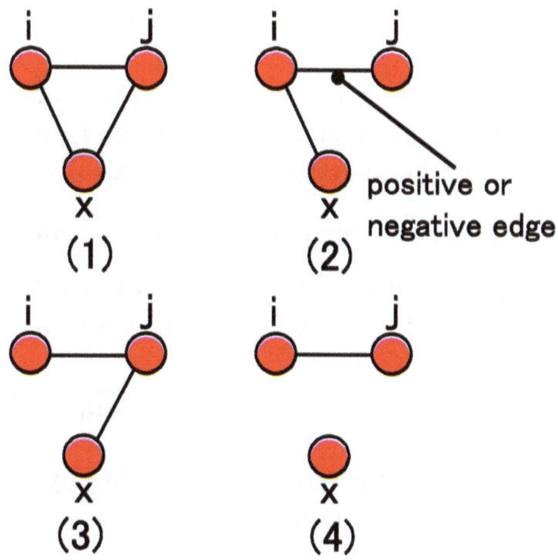

Fig. 3 Four connection patterns

communities i, j and x (Fig. 3), we apply appropriate update of equations for each pattern.

Update of equations for ΔQ^+ are as follows:

1. **If x is connected to both i and j by positive edges,**
   then ΔQ^+_{ix} will be computed as follows:

$$\begin{aligned}\Delta Q^+_{ix} &= 2(e^+_{ix} + e^+_{jx} - (a^+_i + a^+_j)a^+_x) \\ &= 2(e^+_{ix} - a^+_i a^+_x) + 2(e^+_{jx} - a^+_j a^+_x) \\ &= \Delta Q^+_{ix} + \Delta Q^+_{jx} \end{aligned} \quad (18)$$

   In this pattern, we need only a simple addition.
2. **If x is connected to i but not to j by positive edge,**
   then we substitute $e^+_{jx} = 0$ for Eq. (14) and ΔQ^+_{ix} will be computed as follows:

$$\begin{aligned}\Delta Q^+_{ix} &= 2(e^+_{ix} - (a^+_i + a^+_j)a^+_x) \\ &= 2(e^+_{ix} - a^+_i a^+_x) - 2a^+_j a^+_x \\ &= \Delta Q^+_{ix} - 2a^+_j a^+_x \end{aligned} \quad (19)$$

3. **If x is connected to j but not to i by positive edge,**
   then we substitute $e^+_{ix} = 0$ for Eq. (14) and ΔQ^+_{ix} will be computed as follows:

$$\Delta Q^+_{ix} = \Delta Q^+_{jx} - 2a^+_i a^+_x \quad (20)$$

4. **If x is not connected to i, j by positive edge,**
   then we substitute $e^+_{ix} = e^+_{jx} = 0$ for Eq. (14) and ΔQ^+_{ix} will be computed as follows:

$$\Delta Q^+_{ix} = -2(a^+_i + a^+_j)a^+_x \tag{21}$$

This is the case that i, j and x are connected only by negative edges. In this case, there is no positive edge between i, j and x. This is the specific pattern for signed networks and original CNM does not consider this pattern.

Update equations for ΔQ^- are as follows:

1. **If x is connected to both i and j by negative edges,** then ΔQ^-_{ix} will be computed as follows:

$$\Delta Q^-_{ix} = \Delta Q^-_{ix} + \Delta Q^-_{jx} \tag{22}$$

2. **If x is connected to i but not to j by negative edge,** then we substitute $e^-_{jx} = 0$ for Eq. (15) and ΔQ^-_{ix} will be computed as follows:

$$\Delta Q^-_{ix} = \Delta Q^-_{ix} - 2a^-_j a^-_x \tag{23}$$

3. **If x is connected to j but not to i by negative edge,** then we substitute $e^-_{ix} = 0$ for Eq. (15) and ΔQ^-_{ix} will be computed as follows:

$$\Delta Q^-_{ix} = \Delta Q^-_{jx} - 2a^-_i a^-_x \tag{24}$$

4. **If x is not connected to i, j by negative edge,** then we substitute $e^-_{ix} = e^-_{jx} = 0$ for Eq. (15) and ΔQ^-_{ix} will be computed as follows:

$$\Delta Q^-_{ix} = -2(a^-_i + a^-_j)a^-_x \tag{25}$$

This is the case that i, j and x are connected only by positive edges. In this case, there is no negative edge between i, j and x.

These updates are continued until the largest ΔQ_{signed} becomes negative.

Our method reduces computational cost by calculating not the whole Q_{signed} but ΔQ_{signed} that would result from the agglomeration. In addition, it also reduces computational cost by avoiding to calculate e^+_{ij} and e^-_{ij}, which are time consuming. These ingenuities do not affect the resultant value but they significantly improve the efficiency of our method.

5 Experiments

5.1 Synthetic Networks

One way to test our algorithm is to see how well it performs when it is applied to synthetic signed networks. The generated network is composed of 128 nodes which

Table 1 Calculation times

Method	Time (s)
Calculate the whole of Q_{signed} (not ΔQ_{signed}) after agglomeration	202.0
Calculate ΔQ_{signed} without using update rules (Eqs. 18–25)	2.85
Our method	0.35

are split into four communities containing 32 nodes each. We regard these four communities as correct answer communities. The purpose of this experiment is to examine whether the answer communities can be extracted. The generation process is the same as the one used in the experiments by Danon et al. [2]. p_1 is the noise rate from positive to negative, and p_2 is the noise rate from negative to positive.

We detect communities in signed networks ($p_1 = p_2 = 0.05$) by our method. In order to compare the calculation time, we also detect communities with two other methods. The first method is "calculate the whole of Q_{signed} (not ΔQ_{signed}) after agglomeration". The second method is "calculate ΔQ_{signed} without using update rules (Eqs. 18–25)". As a result of these three methods ($\alpha = 0.5$), the four communities are detected correctly. The calculation times are shown in Table 1. Our method is quite faster than other two methods.

In order to examine the impact of p_1 and p_2, we detect communities in signed networks where we set one parameter as 0.05 and change the other from 0.05 to 0.5. 10 signed networks are generated for each state. We use our method where $\alpha =$ 0.2, 0.5, 0.8 and examine the accuracy of detected communities, which is evaluated by NMI (Normalized Mutual Information) [2].

Figure 4 shows the result when p_2 is fixed and p_1 is changed, and Fig. 5 shows the result when p_1 is fixed and p_2 is changed. X-axis is the value of p_1(or p_2), and y-axis is the value of NMI. When α is large, the value of NMI is also large, but when α is small, the value of NMI is small.

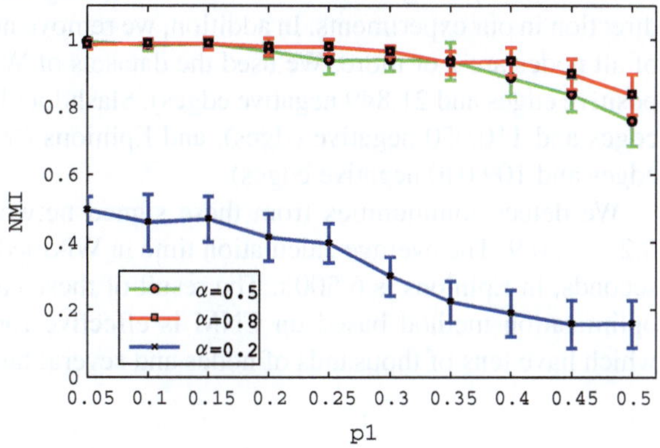

Fig. 4 Results when p_1 is changed

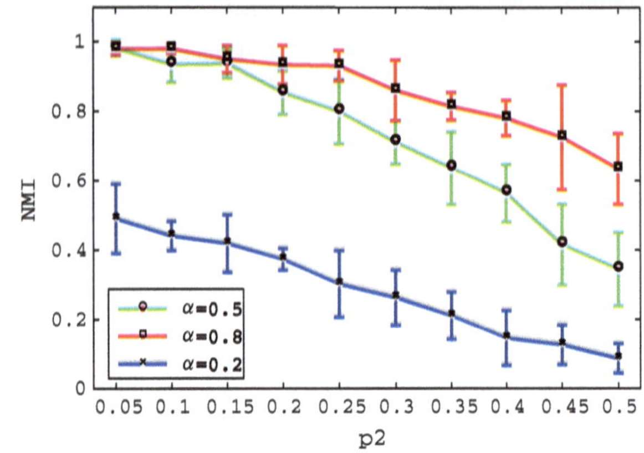

Fig. 5 Results when p_2 is changed

When α is small, the negative edges between communities are more focused. The density of positive edge in the answer community is less important than when α is larger. As a result, it becomes hard to agglomerate a pair within the answer community.

Besides, when p_1 (noise rate within the answer community) is large, the value of NMI is relatively larger than the result when p_2 (noise rate between answer communities) is large. When p_2 is large, the number of positive edges between answer communities increases. Therefore, the chance of agglomeration between answer communities is raised. As a result, detected communities become different from the correct answer, and the value of NMI becomes small.

5.2 *Real-World Networks*

We use three real-world signed networks for our experiments. Each data can be obtained from Stanford Large Scale Network Dataset (http://snap.stanford.edu/data/index.html) [6]. Original networks are directed signed networks, but we ignore edge direction in our experiments. In addition, we remove nodes with degree 1, so degrees of all nodes are 2 or more. We used the datasets of Wikipedia (4,786 nodes, 76,607 positive edges and 21,849 negative edges), Slashdot (47,726 nodes, 329,873 positive edges and 110,050 negative edges), and Epinions (60,332 nodes, 535,303 positive edges and 109,040 negative edges).

We detect communities from these signed networks while changing $\alpha = 0.1, 0.2, \ldots, 0.9$. The average calculation time in Wikipedia is 70 s, in Slashdot is 4,800 seconds, in Epinions is 6,500 s. The result of these calculation times show that our optimization method based on CNM is effective for large-scale signed networks which have tens of thousands of nodes and several hundreds of thousands of edges.

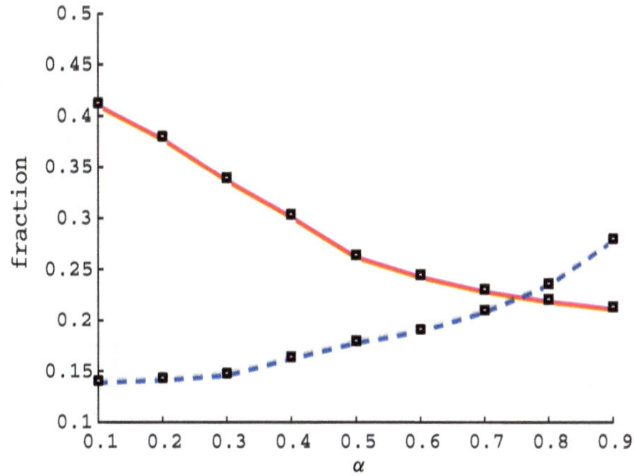

Fig. 6 The value of Eq. (26) (*red line*) and Eq. (27) (*blue dotted line*) in Epinions

The property of positive and negative edges for each α in Epinions is shown. In Fig. 6, the value of Eq. (26) is represented by continuous line.

$$\frac{1}{2m^+}\sum_{ij} A^+_{ij}(1-\delta(c_i, c_j)) \tag{26}$$

Equation (26) shows the fraction of positive edges between communities. Therefore, the result is good for positive edges if this value is small.

In Fig. 6, the value of Eq. (27) is represented by dotted line.

$$\frac{1}{2m^-}\sum_{ij} A^-_{ij}\delta(c_i, c_j) \tag{27}$$

Equation (27) shows the fraction of negative edges within communities. Therefore, the result is good for negative edges if this value is small.

In Fig. 6, when α is small, the number of positive edges between communities is large but within communities is small. On the other hand, when α is large, the number of negative edges within community is large but between communities is small.

In addition, Figs. 7 and 8 are about fraction of positive and negative edges connected to the largest community in each result. X-axis is the value of α. Y-axis of Fig. 7 is fraction of negative edges connected to the largest community and y-axis of Fig. 8 is fraction of positive edges connected to the largest community.

In Fig. 7, when α is small, the largest community tends to gather a lot of negative edges, but when α is large, it does not gather a lot. On the other hand, in Fig. 8, when α is small, the largest community tends to gather a lot of positive edges, but when α is large, it does not gather a lot. Figures 7 and 8 show foes will be placed in different communities, but friends also will be placed in different communities when

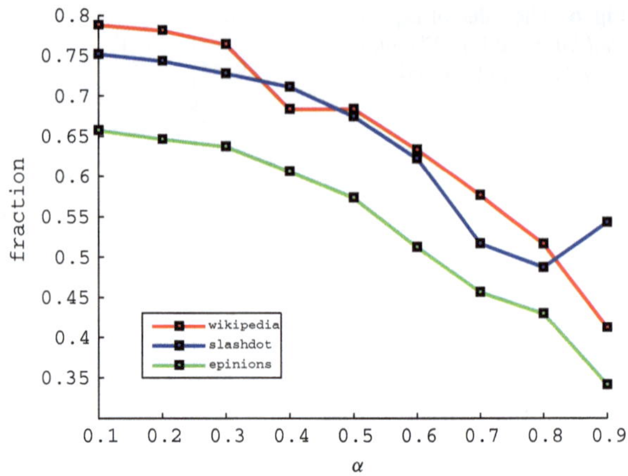

Fig. 7 Fraction of negative edges connected to the largest community

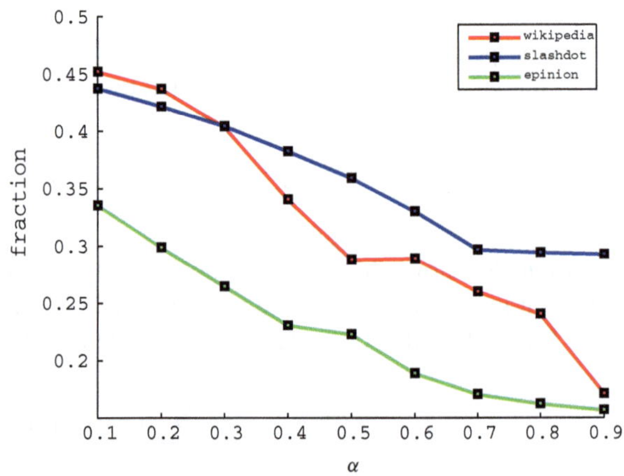

Fig. 8 Fraction of positive edges connected to the largest community

α is small. In contrast, friends will be placed in the same community, but foes also will be placed in the same community.

From these results, when α is small, the number of negative edges within communities is small and between communities is large. It is a good community structure in terms of negative edges. When α is large, the number of positive edges within communities is large and between communities is small. It is a good community structure in terms of positive edges. Certainly, the parameter α we introduce works as our intention.

In modularity by Gomez et al., they fix α to the fraction of positive edges. Generally, in most real world signed networks, the number of positive edges is more than the number of negative edges. Thus, community detection with their modularity will be biased for the result which allows negative edges within communities. According to Traag et al. [12], it is difficult to determine the optimal value of α. It depends on network structure and characteristics required for community structure. Therefore,

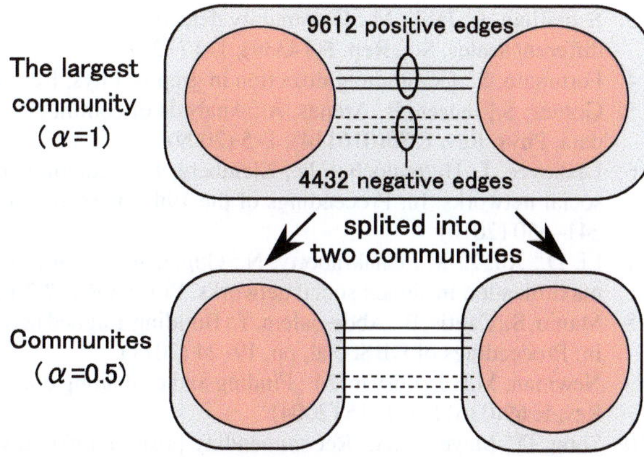

Fig. 9 The largest community with $\alpha = 1$ and communities with $\alpha = 0.5$ (wikipedia)

in order to try several experiments with values of α, the fast optimization method proposed in this paper is important.

We also examine the difference between the result where negative edges are ignored and the results where negative edges are considered. Figure 9 is about the largest community in Wikipedia with $\alpha = 1$ (negative edges are ignored) and communities with $\alpha = 0.5$. Most of nodes in the largest community with $\alpha = 1$ are split into the members of two communities with $\alpha = 0.5$. There are 9,612 positive edges (12.5% of positive edges) and 4,432 negative edges (20.3% of negative edges) between them. Because there are a lot of negative edges between them, they might be hostile each other. However, they are regarded as members of a community when negative edges are ignored. Therefore, both types of edges should be considered appropriately. By adjusting the value of α, we can detect inner factions within communities.

6 Conclusion

We have extended signed modularity and CNM in order to detect communities in large-scale signed networks. We detect communities in synthetic signed networks by our method and examined the relationship between α in our modularity and positive and negative edges in the resultant communities. From the result of real world signed networks, we can say that our method is effective also for large-scale signed networks.

References

1. Clauset, A., Newman, M.E.J., Moore, C.: Finding community structure in very large networks. Phys. Rev. **70**(6) (2004)
2. Danon, L., Diaz-Guilera, A., Duch, J., Arenas, A.: Comparing community structure identification. J. Stat. Mech. **P09008**, 1–10 (2005)

3. Esmailian, P., Jalili, M.: Community detection in signed networks: the role of negative ties in different scales. Sci. Rep. **5**(14339), 1–17 (2015)
4. Fortunato, S.: Community detection in graphs. Phys. Rep. **486**, 75–174 (2010)
5. Gomez, S., Jensen, P., Arenas, A.: Analysis of community structure in networks of correlated data. Phys. Rev. E **80**(016114), 1–5 (2009)
6. Leskovec, J., Huttenlocher, D., Kleinberg, J.: Predicting positive and negative links in online social networks. In: Proceedings of the 19th International World Wide Web Conference, pp. 641–650 (2010)
7. Li, D., Xu, Z.-M., Chakraborty, N., Gupta, A., Sycara, K., Li, S.: Polarity related influence maximization in signed social networks. PLOS ONE **9.7**(e102199), 1–12 (2014)
8. Maniu, S., Cautis, B., Abdessalem, T.: Building a signed network from interactions in wikipedia. In: Proceedings of DBSocial, pp. 19–24 (2011)
9. Newman, M.E.J., Girvan, M.: Finding and evaluating community structure in networks. Phys. Rev. E **69**(026113), 1–15 (2004)
10. Song, D., Meyer, D.A.: Recommending positive links in signed social networks by optimizing a generalized auc. In: Proceedings of the Twenty-Ninth AAAI Conference on Artificial Intelligence, pp. 290–296 (2015)
11. Szell, M., Lambiotte, R., Thurner, S.: Multirelational organization of large-scale social networks in an online world. PNAS **107**(31), 13636–13641 (2010)
12. Traag, V., Bruggeman, J.: Community detection in networks with positive and negative links. Phys. Rev. E **80**(036115) (2009)

Characterising Inter and Intra-Community Interactions in Link Streams Using Temporal Motifs

Jean Creusefond and Remy Cazabet

Abstract The analysis of dynamic networks has received a lot of attention in recent years, thanks to the greater availability of suitable datasets. One way to analyse such dataset is to study temporal motifs in link streams, i.e. sequences of links for which we can assume causality. In this article, we study the relationship between temporal motifs and communities, another important topic of complex networks. Through experiments on several real-world networks, with synthetic and ground truth community partitions, we identify motifs that are overrepresented at the frontier—or inside of—communities.

1 Introduction

Communication networks represent human interactions that happen at certain times. The properties of these networks are often studied in order to have a better understanding of human dynamics [2, 8].

The basic building blocks of networks are called motifs, small structures that appear multiple times in the network. This concept was originally formulated for static networks [12] and has been extended for temporal networks [19]. In the case of communication networks, these motifs are an indication of the nature of the communication [18]. For instance, a set of messages in a back-and-forth pattern between two individuals is probably a conversation.

It is a common assumption that the nature of the relationship of two individuals define the nature of the communities that they share [1]. If the motifs characterise the relationships between individuals, they may be related to the community structure.

The existing definitions of motifs describe messages that are received and sent in a short time-frame. Such motifs do not include causally-linked interactions that

J. Creusefond (✉)
GREYC, Normandie Université, Caen, France
e-mail: jean.creusefond@unicaen.fr

R. Cazabet
Sorbonne Universites, UPMC Univ Paris 06, CNRS, LIP6 UMR,
7606 Paris, France

B. Gonçalves et al. (eds.), *Complex Networks VIII*,
Springer Proceedings in Complexity, DOI 10.1007/978-3-319-54241-6_7

happen outside of the time-frame. These interactions could be due to an individual that is not active on the network at that time, and therefore unaware of the messages received.

In this paper, we first propose an adaptation of the definition of a motif that takes into account users' activity periods. We then study experimentally the frequency of motifs inside and outside communities in order to test the hypothesis that temporal motifs are linked to the community structure.

2 Related Work

Zhao et al. [19] defined temporal motifs. They measured the frequency of the different motifs and characterised them by their shape (ping pong, star, çhain). Kovanen et al. [10] extended the definition of motifs in order to take into account the order of communications. For instance, their definition differentiates a "AB-BA-AB" motif from a "AB-AB-BA" motif, which the previous definition does not.

Zhang et al. [18] considered the relative frequency of some 3-events motifs when increasing the time window. They observed that the dominant 3-event motifs were related to the dominant 4-event motifs in the 6 datasets that were used.

In order to decide of their significance, the frequency of the motifs in the dataset is often compared with null models [16, 19]. These models describe a network that is identical to the data, except for one feature that is randomised. This methodology is used to evaluate the influence of the randomised feature on the measurements.

Zhao et al. [19] compared their results to the time-mixing model, a null model where all timestamps of the dataset are randomised. They observed that the time-mixing model created mainly isolated entries, which is an important difference with empirical observations. However, the time-mixing model deletes the phenomenon of burst in the activity of individuals, on top of deleting causality effects.

Tabourier et al. [16] presented a null model that conserves this feature, the correlation-mixing model. As for the time-mixing model, all source and destinations are kept and timestamps are randomised. However, this randomisation is carried out over the messages that were emitted by the same individual, and not over all messages. It implies that temporal features such as the burstiness of communications is conserved, but not the causal link between messages.

Several works have been done on the question of detecting communities on dynamic networks [4]. However, these approaches focus on slowly evolving networks, in which edges are persistent along time (relations, for instance friendship or colleague relation). On the contrary, this work focuses on networks which have a much faster temporality than communities, i.e. interactions are short-lived (for instance messages, calls between friends or colleagues). We therefore assume a fixed community structure, and observe interactions over this structure.

3 Adapting Motifs for Communication Networks

In this section, we will introduce a variation on motifs that take into account the activity periods of individuals. We call this variation an a-motif.

We model the communication network as links streams $G = (V, E)$. A link stream is composed of a set V of nodes and a set $E \subset V \times V \times \mathbb{R}^+$ of timestamped links between nodes. We note that multiple links may exist between the same pair of nodes.

A temporal motif describes the structure of a sequence of communications. Formally, a temporal motif is an equivalence class of a communication graph [19], that is defined as follows on link streams:

Definition 1 (*communication graph*) A communication graph on a window of size $W \in \mathbb{R}^+$ is a link stream $G = (V, E)$ such that $\forall (u_i, v_i, t_i) \in E$, $\exists (u_j, v_j, t_j) \in E$ that respects $(u_i, v_i, t_i) \neq (u_j, v_j, t_j)$, $\{u_i, v_i\} \cap \{u_j, v_j\} \neq \emptyset$ and $0 < |t_i - t_j| < W$.

Two communication graphs belong to the same equivalence class (i.e. motif) if the corresponding weighted graphs (a link is weighted by the number of communications) are isomorphic. Kovanen et al. [10] extend this equivalence relationship by taking into account the order of the links in the communication graphs. We call a communication graph that belong to such an equivalence class an *instance of a motif.*

This paper focuses on communication networks such as e-mails or answers in an online forum. In such networks, the individual receiving a message is not always aware of the message at the time of reception. Typically, receiving an e-mail does not mean that it is acknowledged. In that case, the causal link between two communications may not be directly related to the reaction time. We define the a-motif (for *activity motifs*) in order to take that phenomenon into account.

We first split the messages emitted by the individuals into activity periods. These periods are time intervals when an individual emits messages in a short burst.

Definition 2 (*μ-activity period*) For each node $v \in V$ in a link stream $G = (V, E)$, we note E_v the set of messages emitted by v and $med(v \in V)$ the median of the time elapsed between two consecutive messages emitted by v. We also note $t((u, v, x) \in E) = x$ the date of an edge. A **μ-activity period** of an individual $v \in V$ is a time interval $[a; b]$ during which v emitted a set of messages $M(a, b) = \{e \in E_v \mid a \leq t(e) \leq b\}$, that respects the following properties:

- $\exists e_1 \in M(a, b), t(e_1) = a$ and $\exists e_2 \in M(a, b), t(e_2) = b$ and
- $\forall e_1 \in M(a, b), t(e_1) \neq b \Rightarrow \exists e_2 \in M(a, b), 0 < t(e_2) - t(e_1) \leq \mu \cdot med(v)$ and
- $\forall e \in E_v,\ t(e) < a \Rightarrow t(e) < a - \mu \cdot med(v)$ and $t(e) > b \Rightarrow t(e) > b + \mu \cdot med(v)$.

We then define the a-motifs as equivalence classes of *activity graphs*, formed as follows. If an edge (u_1, v_1, t_1) belongs to an activity graph, the edge (u_2, v_2, t_2) may also belong in that graph if $t_1 < t_2$ and:

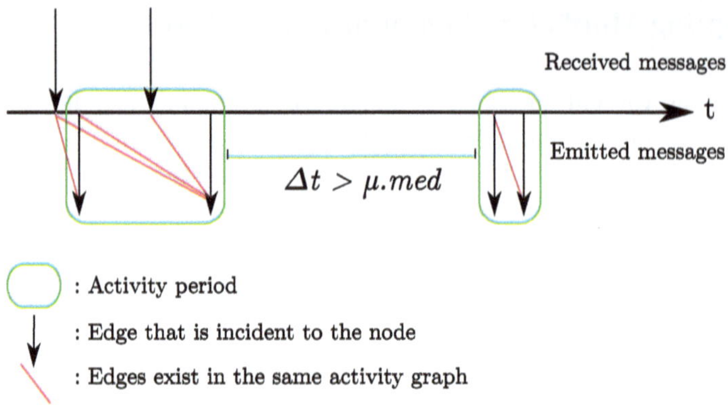

Fig. 1 For a node, the messages that are emitted are grouped into activity periods. The set of incident edges forms activity graphs

- $u_1 = u_2$ and t_1 and t_2 belong to the same activity period of u_1. There might be a causal link between two messages emitted by an individual in the same activity period.
- $v_1 = u_2$ and t_2 belong in the next activity period of u_2 that happens after t_1. If t_1 is inside an activity period of u_2, then t_2 must belong to the same activity period. There might be a causal link between a message received and the next messages sent by the recipient during his/her next activity period.

We use the equivalence function introduced in Kovanen et al. [10] to define the a-motifs as equivalence classes of activity graphs. The detection of a-motifs instances is illustrated Fig. 1.

For complexity reasons, we restrict our study to size 3 a-motifs, i.e. those that are made of three edges. This size is chosen as a compromise between the computation time needed for the detection of instances and the complexity of the structures that are observed.

We identify the motifs by letters that correspond to the nodes that are involved in the motif.

Some size 3 a-motifs are geometrically similar, such as "AB-AC-BC" and "AB-AC-CB", or "AB-BC-CB" and "AB-BA-BC". In order to reduce the number of observations, we focus on four motifs that have been identified as important in the associated literature [16, 18, 19] and a fifth that we identified as interesting. Those are: the star "AB-AC-AD", the ping-pong "AB-BA-AB", the triangle "AB-BC-CA" and the chain "AB-BC-CD". We add the spam "AB-AB-AB" to that list because of its direct possible interpretation. Those motifs are illustrated Fig. 2.

There may be activity periods including dozens of messages while others include only a few. If an activity period of a node v is made of k edges and if v received l messages before that period, then $k \cdot l$ instances of size 2 a-motifs are created. The impact of a message on a-motifs frequency is therefore dependent of the size of the activity periods of the receiver.

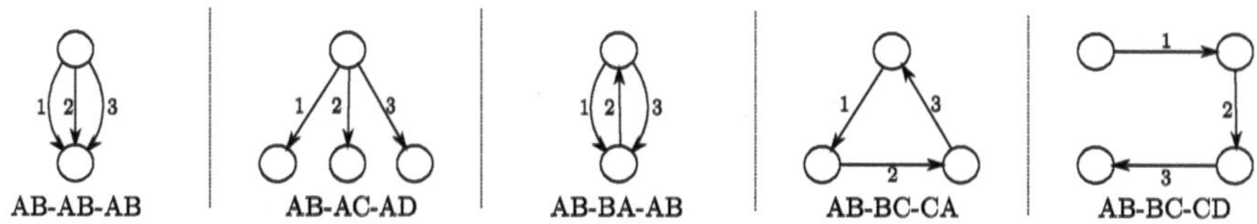

Fig. 2 The five studied motifs. *Numbers* indicate the order of the edges

In this work, we consider that a message should not have more impact on the results than another because of the size of activity periods. To that purpose, we weight instances of a-motifs such that the weights of the set of instances that has the original edge sum to one. That weight is computed in the following manner: from an instance that has a weight w, if that instance is extended to generate k instances of bigger size, each of these instances has a w/k weight.

For instance, if an edge creates k_1 instances of size two, each of them has weight $1/k_1$. If the first of these instances generates k_2 instances of size three, each of them has weight $1/(k_1 \cdot k_2)$. If the second of these instances generates k_2' instances of size three, each of them has a $1/(k_1 \cdot k_2')$ weight, and so on. Each measure that is presented in following experiments is weighted accordingly.

4 Experiments

In this section, we present our study of the properties of a-motifs.

These experiments were implemented in Python. They were run in parallel on 40 AMD Opteron CPUs (2.6 GHz). Due to the size of the dataset and the number of null-model instances, the full run takes about a day.

4.1 Datasets

In order to carry out our experiments, we collected a dataset that includes messages between individuals and three ground-truth community partitions. This dataset is original since, to the best of our knowledge, no openly available dataset features both types of data.

4.1.1 Caen University Dataset

We obtained metadata for all emails transferring through servers of Caen University, France, for a period of 3 months. Available information include source, destination and timestamp. Individuals in this network are students and employees of the university.

Table 1 Konect's networks

Name	n	m	Nodes	Edges
Enron [9]	86978	1134990	Employees	E-mails
Facebook [17]	45813	855542	Users	Wall posts
UC Irvine [13]	1899	59835	Students	Messages
Radoslaw [11]	167	82876	Employees	E-mails
Debian [6]	34648	316569	Users	Answers
Digg [5]	30360	86203	Users	Answers
Linux Kernel mailing list (LKML)[a]	26885	1028233	Users	Answers
Slashdot [7]	51083	139789	Users	Answers

[a]http://www.konect.uni-koblenz.de/networks/lkml-reply

Three kinds of partitions can be extracted from available data:

- For researchers, we know the **research laboratory** they belong to.
- For students and researchers, we also know their **CNU section** (CNU stands for Universities National Council), which indicates to which scientific field they belong to.
- For all users, we know to which **administrative entity** they belong to, typically their school.

This dataset includes 45 **research laboratories**, 146 **CNU sections** and 57 **administrative entities**.

The network has the following properties:

- It contains 7 688 665 messages sent between 210 085 addresses.
- 168 507 messages sent between 918 addresses with a **research laboratory**.
- 378 721 messages sent between 17 275 addresses with a **CNU section**.
- 1 275 662 messages sent between 26 177 addresses with a **administrative entity**.

We created three link streams, one for each partition, that includes only nodes corresponding to individuals present in the corresponding partition, and that includes communication between these nodes.

4.1.2 Other Datasets

Besides the Caen university dataset, we analysed a set of communication networks available on the Konect[1] website (see Table 1). After filtering out self loops and nodes with no links, we considered them as link streams.

Because these datasets do not have a known ground truth partition, we used Louvain [3] and Infomap [15] community detection algorithms on the aggregated

[1]http://www.konect.uni-koblenz.de.

network to generate two reference partitions. The aggregated network contains an edge between a pair of nodes if there is at least one interaction at any point in time between these two nodes in the link stream. Since the results on the partitions of both algorithms are similar, we will only present the results on the partitions obtained with the Louvain algorithm.

4.2 *Comparing with the Correlation-Mixing Model*

For each measure on the motifs, we compare the value on the original graph and the same value on graphs generated by the correlation-mixing model. We consider statistically significant differences to be a consequence of causality, as described by Tabourier et al. [16].

In practice, we observe that these measures are normally distributed. In such a case, we can use the "66-95-99.7 rule" [14], that states that about 66% of normally distributed values are within one standard deviation of the mean, about 95% of them are within two standard deviations and about 99.7% of them are within three standard deviations. Therefore, a value that is further from the mean than three times the standard deviation would have less than 0.3% chance to be generated by the normal distribution. For each measure s on the data, we obtain the average μ_s and the standard deviation σ_s of s on the graphs generated by the null model. We then evaluate the difference between the data and the null model using the z-score:

$$z\text{-}score(s) = \frac{s - \mu_s}{\sigma_s} \tag{1}$$

If the z-score is more than three in absolute value, we conclude that the null model does not explain the value of the measure in the data. Since we use the correlation-mixing model, a significant difference would be caused by the removal of the correlation between messages in the null model.

4.3 *Experimental Properties of A-motifs*

We start by studying the differences between motifs and a-motifs. In order to have enough messages during activity periods, we take $\mu = 2$. Indeed, $\mu = 1$ implies that half of edges finish an activity period since half of the edges are separated by more than the inter-edge time median. In the datasets, $\mu = 1$ implies that these periods include a small amount of edges.

Zhao et al. [19] observed that star and chain motifs are the most common ones. Analysis of the corresponding a-motifs on our datasets confirm this observation in average (Fig. 3), despite a few exceptions for some datasets. Overall, the chain motif represents 16% of all motifs, stars represent 6%, while ping-pong comes third at 3%.

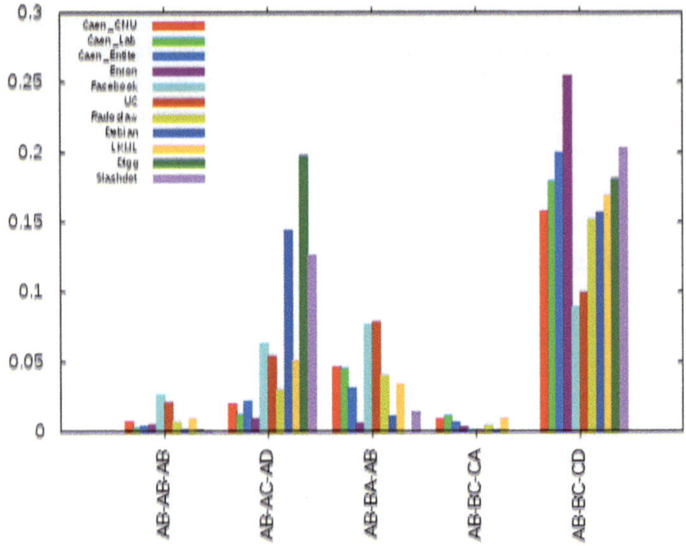

Fig. 3 A-motif frequencies for different networks

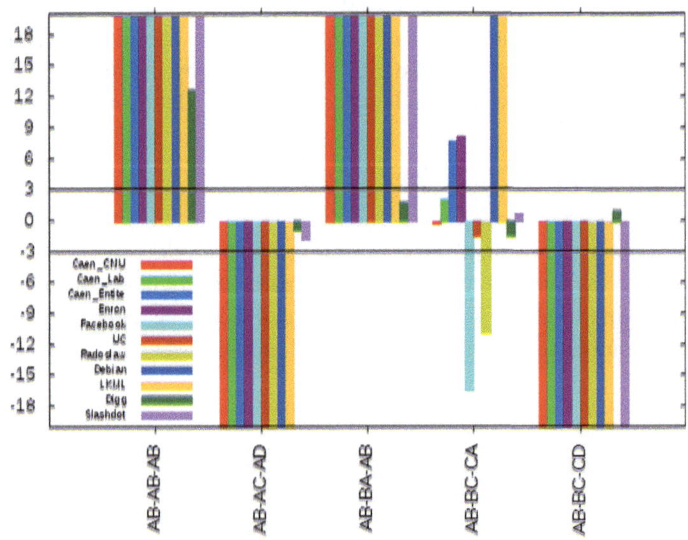

Fig. 4 Z-score of a-motif frequency for different networks. Scores above 3 in absolute value are considered significant. Values beyond 20 are truncated

We also study the z-score of the frequency of each motif Fig. 4. We can observe significant tendencies at least for 4 of the 5 studied motifs: in most networks, stars and chains are less common in observed data than in the null model, while spam and ping-pong are more common.

In [16], a similar analysis was conduced on a phone call dataset, only for stars and chains. While their conclusion for stars was the same than ours, their conclusion for chains was the opposite. This difference might be due to the difference in nature of datasets, or to a difference in the method of analysis: they segmented time using fixed temporal windows, while we used activity periods.

4.4 A-motifs and Communities

In this section, we study the relation between a-motifs and communities. In particular, we are interested to know if some a-motifs are more common inside or in-between communities.

We define the normalised internal weights of a-motifs of type m as:

$$w_{in}^{norm}(m) = \frac{w_{in}(m)}{\sum_{m\prime \in M} w_{in}(m\prime)}$$

with $w_{in}(m)$ the sum of weights of a-motifs of type m that have at least an edge inside a community. We similarly define the normalised external weights.

We now compute a normalised cross-community score for a-motifs of type m:

$$ccscore(m) = \frac{w_{ext}^{norm}(m) - w_{in}^{norm}(m)}{max(w_{ext}^{norm}(m), w_{in}^{norm}(m))}$$

4.4.1 Interpretation of ccscore

This score can vary between -1 and 1, with positive score indicating a higher relative prevalence of cross-community instances, while negative values indicate a-motifs more commonly found inside communities. Results are presented Fig. 5.

We observe that three a-motifs have negative scores in most datasets: spams, ping-pong and triangles. This means that, comparatively to others, these a-motifs tend to occur more inside communities than outside.

The two other a-motifs (star and chain) have less clear tendencies, but seem to occur slightly more often in-between communities.

It is nevertheless important to note that there are notable exceptions to these tendencies, in particular the Caen CNU dataset for spam and chains, or a divergent result for triangles on Digg.

4.4.2 z-Score of ccscore

As previously, we compute the z-score of the ccscore in order to evaluate how significant are the tendencies (see Fig. 6). We observe that most values are significantly

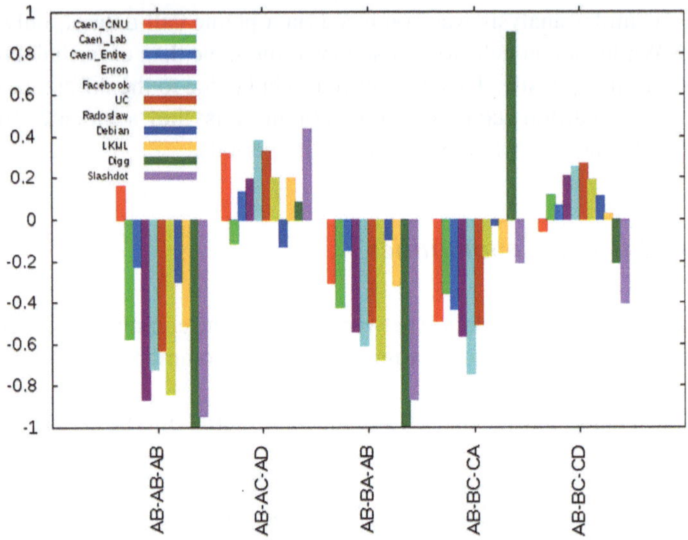

Fig. 5 Ratio between external and internal proportions of a-motifs

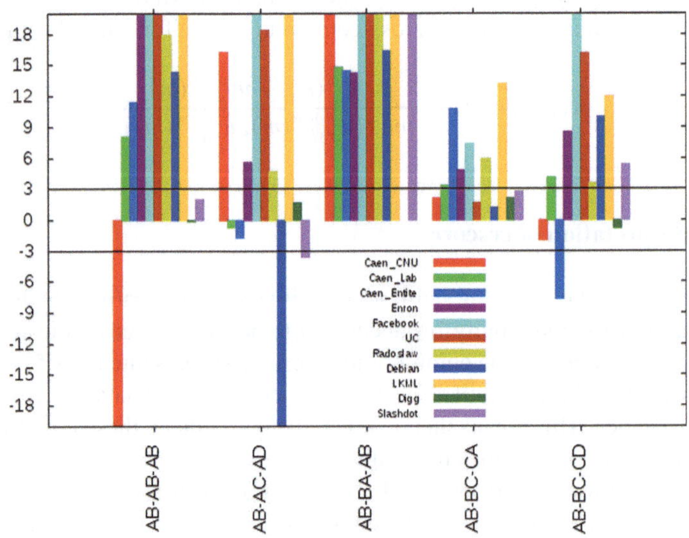

Fig. 6 z-Score of ccscores

higher than those in the null-model, therefore that ccscores observed in the dataset are higher than those in the null model. We conclude that studied a-motifs appear more frequently between communities with respect to the the null-model.

4.5 *Discussion*

In previous sections, we have observed that some a-motifs are more likely to occur inside or outside communities, and that these patterns are significant. As a consequence, we propose that a-motifs could be used, given a temporal network dataset, to distinguish internal and external edges. Identifying such edges could be used to later identify communities.

Another observation is that a-motifs occurring more frequently inside communities seem to be different in nature from those occurring outside. On one hand, inter-community edges are marked by patterns of diffusion of information, including various, different actors: chains and stars. On the other hand, motifs observed inside communities are characterised by an information travelling inside a same set of actors, either several times the same pair of actors (spam, ping-pong), or a cycle coming back to its origin (triangle).

Finally, it is interesting to observe that results are coherent between datasets with ground truth communities (Caen-university) and those in which topological communities have been discovered using the Louvain algorithm. It implies that observed temporal properties are characteristics of structural communities.

5 Conclusion

In this paper, we present an alternate definition of temporal motifs that takes into account the activity periods in communication networks. We measure a large difference of the frequency of these motifs between the empirical data and a null model that ignores causality. This result suggests that our definition captures causally-linked communications.

We also studied the relationship between temporal motifs and community structure. We observed that the conversational motifs such as spam, ping-pong and triangle are generally more frequent inside communities than outside. The star motif, on the other hand, appears more frequently outside communities. The comparison with the null model shows that causally-linked motifs happen frequently outside communities.

These results open the way for future works: on the one hand, it could be possible to detect communities in link streams based on the frequency of a-motifs, taking advantage of our observations. On the other hand, a more detailed analysis of the nature of interactions occurring inside a-motifs could help us to understand better why some of them occur more often inside or outside communities, hence improving the global understanding of the structure of communications.

Acknowledgements This work is funded in part by the European Commission H2020 FET-PROACT 2016-2017 program under grant 732942 (ODYCCEUS), by the ANR (French National Agency of Research) under grants ANR-15-CE38-0001 (AlgoDiv) and ANR-13-CORD-0017-01 (CODDDE), by the French program "PIA—Usages, services et contenus innovants" under grant O18062-44430 (REQUEST), and by the Ile-de-France program FUI21 under grant 16010629 (iTRAC).

References

1. Ahn, Y.-Y., Bagrow, J.P., Lehmann, S.: Link communities reveal multiscale complexity in networks. Nature **466**(7307), 761–764 (2010)
2. Barabasi, A.-L.: The origin of bursts and heavy tails in human dynamics. Nature **435**(7039), 207–211 (2005)
3. Blondel, V.D., Guillaume, J.-L., Lambiotte, R., Lefebvre, E.: Fast unfolding of communities in large networks. J. Stat. Mech. Theory Exp. **2008**(10), P10008 (2008)
4. Cazabet, R., Amblard, F.: Dynamic community detection. In: Encyclopedia of Social Network Analysis and Mining, pp. 404–414. Springer, New York (2014)
5. De Choudhury, M., Sundaram, H., John, A., Seligmann, D.D.: Social synchrony: predicting mimicry of user actions in online social media. In: CSE'09. International Conference on Computational Science and Engineering, 2009, vol. 4, pp. 151–158. IEEE (2009)
6. Gaumont, N., Viard, T., Fournier-Sniehotta, R., Wang, Q., Latapy, M.: Analysis of the temporal and structural features of threads in a mailing-list. In: Complex Networks VII, pp. 107–118. Springer (2016)
7. Gmez, V., Kaltenbrunner, A., Lpez, V.: Statistical analysis of the social network and discussion threads in slashdot. In: Proceedings of the 17th International Conference on World Wide Web, pp. 645–654. ACM (2008)
8. Karsai, M., Kivel, M., Pan, R.K., Kaski, K., Kertsz, J., Barabsi, A.-L., Saramki,. J.: Small but slow world: how network topology and burstiness slow down spreading. Phys. Rev. E **83**(2) (2011)
9. Klimt, B., Yang, Y.: Introducing the Enron Corpus. In: CEAS (2004)
10. Kovanen, L., Karsai, M., Kaski, K., Kertsz, J., Saramki, J.: Temporal motifs. In: Temporal Networks, Understanding Complex Systems. Springer, Heidelberg (2013)
11. Michalski, R., Palus, S., Kazienko, P.: Matching organizational structure and social network extracted from email communication. Business Information Systems, vol. 87, pp. 197–206. Springer, Heidelberg (2011)
12. Milo, R.: Network motifs: simple building blocks of complex networks. Science **298**(5594), 824–827 (2002). Oct
13. Opsahl, T., Panzarasa, P.: Clustering in weighted networks. Soc. Netw. **31**(2), 155–163 (2009). May
14. Pukelsheim, F.: The three sigma rule. Am. Stat. **48**(2), 88–91 (1994)
15. Rosvall, M., Bergstrom, C.T.: Maps of random walks on complex networks reveal community structure. Proc. Natl. Acad. Sci. **105**(4), 1118–1123 (2008)
16. Tabourier, L., Stoica, A., Peruani, F.: How to detect causality effects on large dynamical communication networks: a case study. In: 2012 Fourth International Conference On Communication Systems and Networks (COMSNETS), pp. 1–7. IEEE (2012)
17. Viswanath, B., Mislove, A., Cha, M., Gummadi, K.P.: On the evolution of user interaction in facebook. In: Proceedings of the 2nd ACM Workshop on Online Social Networks, pp. 37–42. ACM (2009)
18. Zhang, Yi-Qing, Li, Xiang, Jian, Xu, Vasilakos, A.: Human interactive patterns in temporal networks. IEEE Trans. Syst. Man Cybern. Syst. **45**(2), 214–222 (2015)
19. Zhao, Q., Tian, Y., He, Q., Oliver, N., Jin, R., Lee, W.-C.: Communication motifs: a tool to characterize social communications. In: Proceedings of the 19th ACM International Conference on Information and Knowledge Management, pp. 1645–1648. ACM (2010)

Part III
Dynamics of Networks

Modeling the Impact of Privacy on Information Diffusion in Social Networks

Livio Bioglio and Ruggero G. Pensa

Abstract Humans like to disseminate ideas and news, as proved by the huge success of online social networking platforms such as Facebook or Twitter. On the other hand, these platforms have emphasized the dark side of information spreading, such as the diffusion of private facts and rumors in the society. Fortunately, in some cases, online social network users can set a level of privacy and decide to whom to show their information. However, they cannot control how their friends will use this information. The behavior of each user depends on her attitude toward privacy, that has a crucial role in the way information propagates across the network. With the aim of providing a mathematical tool for measuring the exposure of networks to privacy leakage risks, we extend the classic Susceptible-Infectious-Recovered (SIR) epidemic model in order to take the privacy attitude of users into account. We leverage such model to measure the contribution of the privacy attitude of each individual to the robustness of the whole network to the spread of personal information, depending on its structure and degree distribution. We study experimentally our model by means of stochastic simulations on four synthetic networks generated with classical algorithms.

Keywords Complex networks · Modeling · Information diffusion · Privacy

1 Introduction

Humans are social animals that love to disseminate ideas and news, as proved by the huge success of social networking websites such as Facebook or Twitter. On the other hand, these platforms have emphasized the dark side of information spreading such as the diffusion of private facts and rumors that may additionally foster slander and cyberbullying acts [21]. As a consequence, the users of online social networks are acquiring a new awareness of the importance of their own privacy on the Web.

L. Bioglio (✉) · R.G. Pensa
Department of Computer Science, University of Turin, Turin, Italy
e-mail: livio.bioglio@unito.it

R.G. Pensa
e-mail: ruggero.pensa@unito.it

B. Gonçalves et al. (eds.), *Complex Networks VIII*,
Springer Proceedings in Complexity, DOI 10.1007/978-3-319-54241-6_8

However, although most users do not disclose very sensitive facts (private life events, diseases, political ideas, sexual preferences, and so on), they are simply not aware of the risks due to the disclosure of less sensitive information, such as GPS tags, photos taken during a vacation period, page likes, or comments on news. Some social media provide advanced tools for controlling the privacy settings of the user's profile [26]. However, yet a large part of Facebook content is shared with the default privacy settings and exposed to more users than expected [17]. According to Facebook CTO Bret Taylor, even though most people have modified their privacy settings,[1] in 2012, still "13 million users (in the United States) said they had never set, or didn't know about, Facebook's privacy tools[2]". Moreover, even though the users of these social networks can usually set a level of privacy, and specify which of their contacts are allowed to see their notifications, they do not have any control on how these contacts will use the information: friends could spread the rumor through other social networks, blogs, websites, medias or simply with face-to-face communication.

The behavior of an individual in these situations highly depends on her level of privacy awareness: an aware user tends not to share her private information, or the private information of her friends on social networks, while an unaware user could not recognize an information as private, and could share it without care to her contacts, even to untrusted ones, putting at risk her privacy or the privacy of her friends. Users' privacy awareness then turns into the so-called "privacy attitude", i.e., the users' willingness to disclose their own personal data to other users, that can be measured by leveraging the way users customize their privacy settings in social networking platforms [16, 24].

The privacy attitude of each actor in a social network heavily influences the effects of information propagation, not only for posts that are clearly private [30]. In fact, it is a well-known fact that by leveraging Facebook user's activity (such as "Likes" to posts or fan pages) it is possible to "guess" some very private traits of the user's personality [15]. For instance, a public comment on news posts may reveal the political ideas of the individual. However, the privacy attitude alone is not a good measure of the user's objective privacy leakage, since the latter depends also on other users' attitude to privacy and the way they contribute in the information propagation process. With the aim of providing a mathematical tool for measuring the exposure of networks to privacy leakage risks, in this paper we study the effects of privacy attitude on information propagation by extending the classic Susceptible-Infectious-Recovered (SIR) epidemic model. In this model, an individual may be susceptible, infectious or recovered: a susceptible individual in contact with an infectious one can become infectious with a transmission probability, while an infectious individual naturally recovers from infection with a recovery rate, turning into a recovered individual. The SIR model can be adopted for modeling the spread of information in a social network [12]: susceptible individuals do not know the information, then are susceptible to be informed; infectious individuals know and spread the information, while recovered individuals already know the information but do not spread it anymore. We extend

[1] http://www.zdnet.com/article/facebook-cto-most-people-have-modified-their-privacy-settings/.
[2] http://www.consumerreports.org/cro/magazine/2012/06/facebook-your-privacy/index.htm.

this compartmental model in order to represent privacy attitude. In our model, each individual belongs to a privacy attitude class that tunes the parameters of the model. The privacy attitude of users has an influence on the way information spreads across the network that additionally unveil its realistic robustness to information leakage as the effects of information propagation within this model. We use our model, by means of stochastic simulations, for studying the role of privacy on the information diffusion in several synthetic networks, generated from classic algorithms, with different distributions of attitude on privacy of their nodes.

The remainder of the paper is organized as follows: we briefly review the related literature in Sect. 2; the privacy-aware propagation model is presented in Sect. 3; Sect. 4 provides the report of our experimental research; Sect. 5 shows how to infer the privacy attitude of a social network user from her profile settings; finally, we draw some conclusions in Sect. 6.

2 Related Work

In epidemiology, the Susceptible-Infectious-Recovered (SIR) epidemic model [13] is employed for modeling infectious diseases that confer lifelong (or long-term) immunity, such as measles, rubella or chickenpox. In this model a susceptible node can become infected, because of the presence of infectious nodes, and an infectious node can naturally recover after few time, gaining immunity to the disease.

The SIR model has been applied to information spreading since early years, even if these applications slightly differ from the common model: in [9] when a spreader meets another infectious node, that already knows the rumor, both lose interest in spreading it any further, and become recovered, while in [18] when two infectious nodes meet, only one node turns into recovered, and the other one remains unchanged. This last version of SIR model for rumor spreading has been widely studied: in [22] the author found that in a complete random network, i.e., a homogeneous network, a rumor can only spread to around the 80% of the total population; more recently in [27] it has been calculated that such percentage is lower than 80% in small-world networks. In [29] the authors found that the number of nodes reached by the rumor depends on the topological structure of the network, decreasing when it changes from random to scale-free network, and on the mean degree of the network, increasing when the mean degree increases; the same happens for the probability of a single node to be informed, that increases when the degree of node increases. Such behavior happens because large hubs are rapidly reached by the rumor, but they easily turn into recovered, preventing the spreading of the rumor to their huge neighborhood. This is confirmed by the observation, in [19], that the density of susceptible nodes at the end of the process decays exponentially with the value of their degree. An extension of this model also allows spontaneous recovery, justified as forgetting mechanism: an infectious node should also turn into recovered spontaneously after a random time. In this case, the model behave more similarly to the classical SIR model, as observed in [20].

In our work, we focus on rumor spreading in presence of a sort of "immunization parameter" that models the privacy attitude of users, i.e., their willingness to disclose their own personal data to other users directly or indirectly. At the best of our knowledge, this is the first attempt of modeling and measuring the robustness of networks to privacy leakage risks by means of a classic epidemic models in social networks. Indeed, a large part of research works on privacy issues in online social networks focus on the anonymization of networked data [28]. Differently from those studies, our work can be positioned in another branch of research that focuses on modeling, measuring and preventing privacy leakage in online social networks. In this regard, one of the most prominent work is [16] where Liu and Terzi propose a framework to compute a privacy score measuring the users' potential risk caused by their participation in the network. This score takes into account the sensitivity and the visibility of the disclosed information and leverages the item response theory as theoretical basis for the mathematical formulation of the score. In [24], the authors define a privacy index that leverages the privacy settings of users to measure their privacy exposure in an online social network according to predefined sensitivity values for users' items. Becker and Chen [7] presents a tool to detect unintended information loss in online social networks by quantifying the privacy risk attributed to friend relationships in Facebook. The authors show that a majority of users' personal attributes can be inferred from social circles. In [23] the authors measure the inference probability of sensitive attributes from friendship links. In [2, 3], the authors define a measure of how much it might be risky to have interactions with them, in terms of disclosure of private information. Among all these research contributions, [16] is the only one that also consider the privacy attitude of users in disclosing their personal data and provide a mathematical formulation for it. This formal definition can be used to tune our information-propagation model according to the attitude towards privacy of the users involved in the social network.

3 A Privacy-Aware Model for Information Spreading

In this section, we introduce the Susceptible-Infectious-Recovered (SIR) epidemic model for modeling the contribution of privacy on information spreading in a social network. Before providing the details of our privacy-aware information-propagation model, we introduce the notation required to formalize the problem.

We consider a social graph G involving a set of n vertices $\{v_1, \ldots, v_n\}$ that are the users participating in G. In this work, the social network is then a represented as a directed graph $G(V, E)$, where V is a set of n vertices and E is a set of directed edges $E = \{(v_i, v_j)\}$. Given a pair of vertices $v_i, v_j \in U$, $(v_i, v_j) \in E$ iif there exists a link from v_i to v_j (e.g., users v_i is in the friend list/circle of v_j or v_j follows v_i). For any given vertex $v_i \in V$ we define the neighborhood $\mathcal{N}(v_i)$ as the set of vertices v_k which vertex v_i is directly connected to, i.e., $\mathcal{N}(v_i) = \{v_k \in V \mid (v_i, v_k) \in E\}$. Conversationally speaking, $\mathcal{N}(v_i)$ is the set of followers of user v_i. Furthermore, we assume that each user v_i belongs to a privacy class $p \in P$, which is defined as the

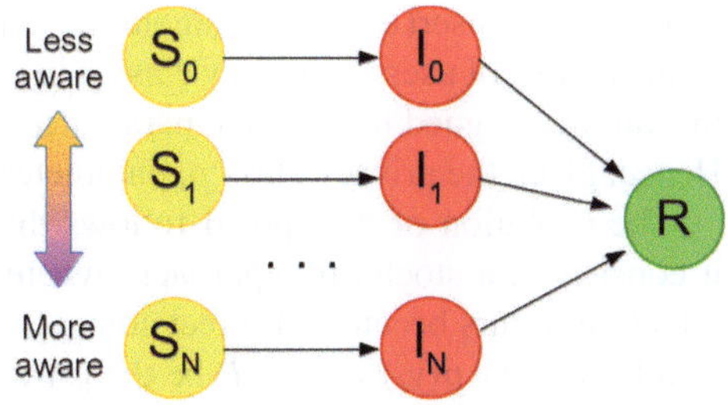

Fig. 1 Transmission model. Each index of compartments S and I represents a privacy class

propensity of an user of the class to disclose her own or other's personal information, directly or indirectly. In practical terms, in online social networks (such as Facebook, Twitter, Instagram or Google+) the privacy class may be unveiled by the way users configure their privacy settings, or the way they post or share/comment other users' posts.

3.1 Information Spreading Model

In the SIR model, at any time step an individual v_i belongs to one compartment among susceptible (S), infectious (I) and recovered (R). An infectious (I) individual v_i may spontaneously recovers from infection with a probability μ, called recovery probability, entering the recovered (R) compartment, or it may spread the disease to a susceptible (S) individual with which it is in contact with a probability λ, called infection probability: the infected susceptible (S) individual immediately becomes infectious (I). We denote with $c(v_i, t) \in \{S, I, R\}$ the compartment of user v_i at time t.

The SIR model can be also applied for the spread of information in a population: susceptible individuals are those who not already know the information, and then they are susceptible to be informed; infectious individuals know the information and actively spread it; finally, recovered individuals are the ones who know the information but do not spread it anymore. The recovery process models a mechanism of aging of the information, that after few time loses its interest or its novelty for an individual and stops to be spread by him. In our formulation, the population is the set of n users $V = \{v_1, \ldots, v_n\}$, while the information may only spread from a user v_i to a user v_j if there exists an edge (v_i, v_j) connecting v_i to v_j.[3]

Here we propose an extension of this model that takes into account the explicit or implicit privacy policies of individuals during the spread of information. A set of privacy classes $P = \{p_0, p_1, \ldots, p_N\}$ is assigned to Susceptible and Infectious compartments, representing the privacy class of an individual belonging to the compartment, and consequently her behavior on information spreading, from less aware (p_0) to more aware (p_N). A graphic representation of our model is given in Fig. 1.

[3]Thus, in our model, the edges are directed from the source of the information to its target.

Moreover we insert a novel parameter $\beta_p \in [0, 1]$ to the SIR transmission model, that is the interest of users in privacy class p in information. Each privacy class differs for the values assigned to the three parameters (β, λ and μ) of the transmission model. Hence, given the privacy class p, parameters β_p, λ_p and μ_p are completely defined.

The evolution of the spread follows the Reed-Frost chain-binomial model [1]: it consists in a stochastic approach, where time is measured in discrete units and infection occurs because of direct contacts. The evolution probabilities are obtained as follows. Let $p(v_i) = p \in P$ be the privacy class of an individual v_i. If it belongs to the susceptible compartment, it may be infected at time $t+1$ with probability:

$$P_{inf}(v_i, t+1) = \beta_p \cdot (1 - \prod_{p' \in P} (1 - \lambda_{p'})^{n_I(v_j,t)}) \quad (1)$$

where $n_I(v_j, t) = |\{v_j \in \mathcal{N}(v_i) \mid c(v_j, t) = I \ \wedge \ p(v_j) = p'\}|$ is the number of individuals in compartment I (infectious) and privacy class p' at time t among the neighbors of individual v_i. Otherwise, if the individual v_i of privacy class p belongs to the infectious compartment I at time t, it may recover with probability μ_p at time $t+1$.

4 Experiments and Results

In this section we provide the results of our experiments performed over several types of synthetic networks. In a nutshell, we generate four networks, each one with a different structure and degree distribution. On each one, we observe the number of nodes reached by the information for three different assignments of privacy classes to the nodes, representing the global attitude on privacy of the network.

4.1 Contact Networks

The information spreads on a contact network, in which nodes represent individuals, and edges between nodes represent contacts between two individuals. Since our objective is to study and characterize the dynamic behavior of the model, here we employ four types of networks, generated with standard algorithms. In all networks, the links between nodes are always considered as reciprocal, i.e., all the graph considered in these experiments are undirected.

The four synthetic networks have approximately the same number of nodes, 75,000, and the same number of edges, around 2,700,000. The first synthetic network is a random graph, also known as an Erdös-Rényi graph [10], generated by means of the fast algorithm in [6]. The second one is a scale-free graph generated with the Barabasi-Albert algorithm [4] where new nodes are attached with 36 edges to

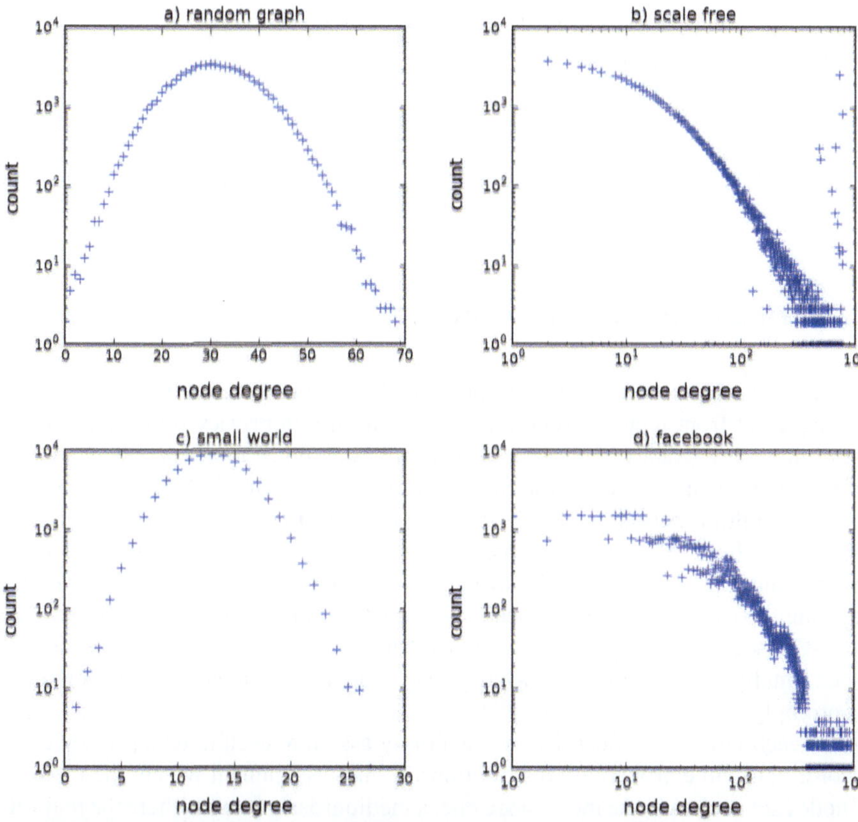

Fig. 2 Degree distribution for each synthetic network

existing nodes with high degree. The third one is a small-world network generated through the Watts-Strogatz mechanism [25] where each node is joined with its 72 nearest neighbors in a ring topology, and each edge has a probability of rewiring equal to 0.15. The fourth one is a Facebook-like network generated using LDBC–SNB Data Generator[4] which produces graphs that mimic the characteristics of real Facebook networks [11]: in particular, we generate a network with 80,000 nodes, but here we consider only the greatest connected component of such network. The degree distributions of these networks are given in Fig. 2.

[4]https://github.com/ldbc/ldbc_snb_datagen.

Table 1 Values of the parameters for the three privacy classes

Parameter	Classes		
	0	1	2
β	0.9	0.5	0.1
μ	0.1	0.3	0.5
λ	0.9	0.5	0.1

4.2 Privacy Class Distributions

In our experiments, we select three privacy classes, numbered from 0 to 2, representing users from unaware (class 0) to more aware on privacy (class 2), in order to provide a few grades of awareness. The values assigned to the parameters of the information spreading model for each class are reported in Table 1. Users in class 0 have a high probability of being interested in information and spreading it over the network for a long period of time ($1/\mu$ is the average duration of the infection). On the contrary, users in class 2 have a very low probability of being interested in information: even if they are reached by information, they spread it only for few time steps. Consequently, the probability of diffusing the information is very low for such users. Finally, class 1 represents average users, then its parameters have been tuned accordingly.

For each network in Sect. 4.1, we randomly assign to each node a privacy class, according to three probability distributions: a safer assignment, where the majority of nodes are in class 2, the most aware one; a medium assignment, where the majority of nodes are in class 1; an unsafer assignment, where the majority of nodes are in the less aware class 0. The number of nodes in each privacy class of these three class distributions are graphically summarized in Fig. 3.

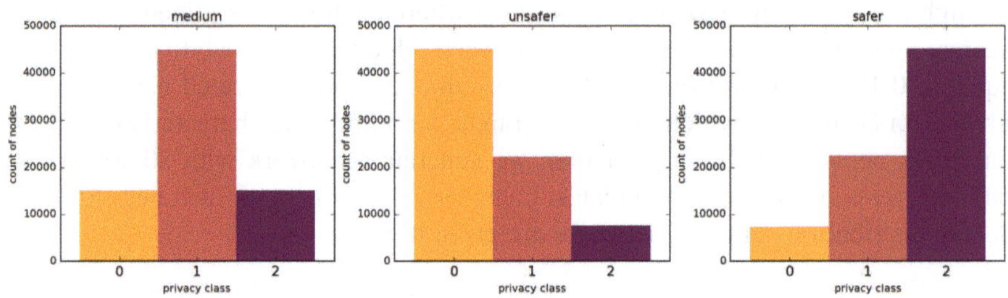

Fig. 3 Class distribution in the three kinds of class assignments

4.3 Experimental Settings

Our experiments are conducted as follows. For each contact network in Sect. 4.1, and for each class distribution in Sect. 4.2, we perform 100 stochastic simulations of information spreading on a completely susceptible population, except for one infectious node. These simulations are repeated for 9 different initial spreaders, randomly chosen among all the nodes, 3 for each privacy class. For each set of simulations we observe the number of informed individuals over time, that is the number of nodes in compartments infectious or recovered, and we calculate the proportion of cases at each time step (prevalence) and the proportion of new cases at each time step (incidence). The results on the same network and class distribution are aggregated.

4.4 Results

We study the impact of the distribution of individuals having different awareness on privacy on the spread of an information in all the synthetic networks described in Sect. 4.1. Figure 4 shows our results for the scale-free network. From the curves of prevalence in Fig. 4a we can notice that the number of informed individuals over time greatly depends on the distribution of privacy classes of the network: where the majority of node is unaware, the information immediately spreads over almost the entire population, while where the network is full of aware individuals the information spreads slowly, and reaches a smaller part of the population. The speed of diffusion is more evident in the curves of incidences, in Fig. 4b, which depicts the proportion of new cases of informed individuals in each time step: under the least safe distribution, the information immediately reaches more than half of population, while for safer distributions this peak is lower, and it is reached few steps later.

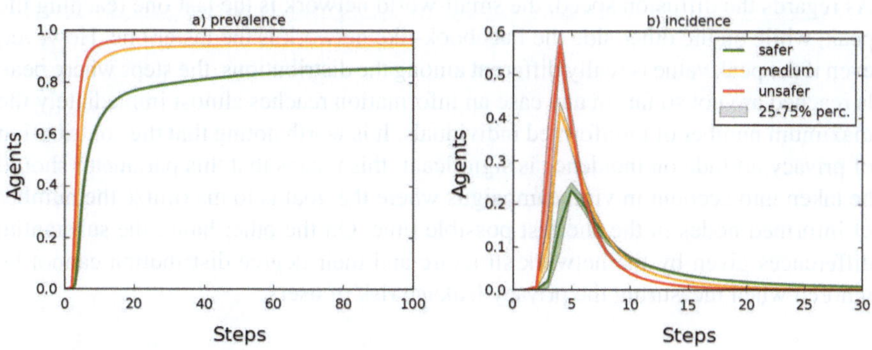

Fig. 4 Prevalence and incidence of informed individuals (ratio) in the scale-free network for each class distribution

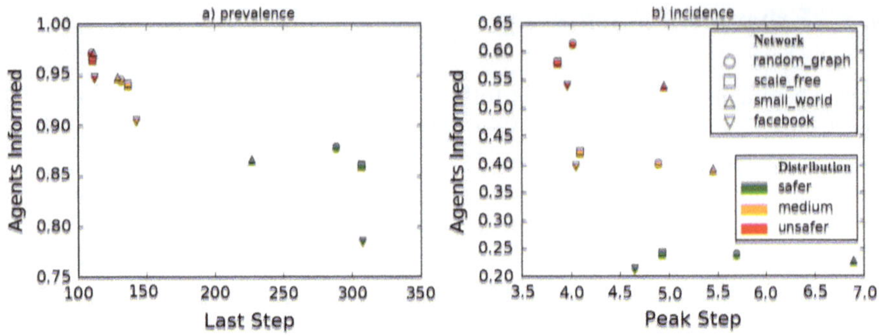

Fig. 5 Prevalence and incidence of informed individuals (ratio) in each network model and class distribution

In order to compare the behavior of all networks in Sect. 4.1, we collect some key features of the prevalence and the incidence curves: for the prevalence ones, we collect the proportion of informed individuals at the end of simulations, that is when there are no more infectious individuals who can spread information, and the step where simulation ends, in order to obtain the duration of the spread and its diffusion among the population; for the incidence curves, we collect the information on the peak of new cases of informed individuals, and the step where the peak is reached, for obtaining a snapshot of the speed of the diffusion of information. These data for all the networks are graphically summarized in Fig. 5.

We can notice that the behavior observed for random graph network happens similarly for all the other networks. Under the safest class distribution, the information reaches a smaller proportion of the population. Furthermore, it stops to be diffused much later than in less safer distributions. Interestingly, even in case of safer distribution an information reaches a huge portion of the population, and such proportion is always smaller for the Facebook-like network: apparently such kind of network is the worst one for spreading an information, especially in case of safer class distribution. As regards the diffusion speed, the small-world network is the last one reaching the peak, while on the other side the Facebook-like network is the faster one. However, even if the peak value is really different among the distributions, the steps where peak is reached are not so far: in any case an information reaches almost immediately the maximum number of uninformed individuals. It is worth noting that the contribution of privacy attitude on incidence is significant: this means that this parameter should be taken into account in viral campaigns where the goal is to maximize the number of informed nodes in the shortest possible time. On the other hand, the substantial differences given by the network structure and their degree distribution cannot be ignored when measuring the privacy leakage risk of users.

5 Privacy Attitude Estimation

In Sect. 3 we have created privacy classes, tuning the characteristic parameters of propagation model, according to the privacy attitude of users. Such attitude, however, involves several psychological, cultural and contextual factors, and it may be indeed difficult to model in real cybersocial systems. In this section we briefly show how to infer it for generic users using some information about their profile settings or disclosing behavior.[5] Our attitude estimation, inspired by the framework defined by Liu and Terzi [16], measures the user's potential risk caused by her participation in the network by assigning to each user a privacy score according to her privacy settings. A $n \times m$ response matrix $\boldsymbol{R}$ is associated to the set of n users and a set of m profile items (e.g., age, gender, education, political views, and so on). Each element r_{ij} of $\boldsymbol{R}$ contains a privacy level that determines the willingness of user i to disclose information associated with profile item j. In [16], the Item Response Theory (IRT) model is adopted to measure the privacy attitude of the users, the sensitivity of the questions, and the probability of a user deciding a given level of visibility to a given profile property. In a binomial case, the probability that a user i sets item j visible to everyone is computed as:

$$P_{ij} = Prob\{r_{ij} = 1\} = \frac{1}{1 + e^{-\alpha_j(\theta_i - \sigma_j)}} \tag{2}$$

where α_j is the discrimination power of item j, σ_j is the sensitivity of j and θ_i is the privacy attitude of user i. In [16], the authors provide an Expectation-Maximization algorithm to estimate parameters α_j and σ_j by only leveraging the response matrix $\boldsymbol{R}$.

When parameters σ_j and α_j ($\forall j \in \{1 \dots m\}$) are known, each θ_i can be computed by maximizing the following log-likelihood function:

$$L = \sum_{j=1}^{m} \left[r_{ij} \log P_{ij} + (1 - r_{ij}) \log (1 - P_{ij}) \right] \tag{3}$$

derived from the likelihood $\prod_{k=1}^{m} P_{ij}^{r_{ij}} (1 - P_{ij})^{1-r_{ij}}$. The solutions can be computed using the Newton-Raphson method, an iterative algorithm that estimates the value of θ_i at iteration t starting from the value of θ_i at iteration $t - 1$ [16].

6 Conclusions

In this paper we have proposed an information propagation model that considers the role of privacy awareness on information spreading inspired by the classical SIR epidemic model. We have assigned different privacy classes to the nodes of

[5] Such information is known by social network providers.

networks, depending on their attitude on privacy, in order to model populations more or less interested on diffusing an information. Through stochastic simulations we have studied the impact of the attitude on privacy of a connected population on the proportion of individuals reached by an information diffused by an unique spreader on a random, a scale-free, a small-world and Facebook-like network.

Our results show that the attitude on privacy can really have an impact on the diffusion of an information, by reducing or increasing the portion of population which receives the information according to safer or less aware attitude on privacy of the individuals on the network. The same behavior happens in all the structures under study, but the Facebook-like network seems to be the most robust to information diffusion.

Our study shows the importance of considering privacy attitude of users in modeling the spreading of rumors, with direct and indirect implications on all applications that involve the dynamics of information spreading, such as influence maximization [14] and community detection [5], as well as on privacy enforcement models and techniques for online social networks, thus inspiring the design of privacy-preserving social networking components for *Privacy by Design* compliant software [8].

Acknowledgements This work was supported by Fondazione CRT (grant number 2015-1638).

References

1. Abbey, H.: An examination of the reed-frost theory of epidemics. Hum. Biol. **24**(3), 201 (1952)
2. Akcora, C.G., Carminati, B., Ferrari, E.: Privacy in social networks: how risky is your social graph? In: Proceedings of IEEE ICDE 2012, pp. 9–19. IEEE Computer Society (2012)
3. Akcora, C.G., Carminati, B., Ferrari, E.: Risks of friendships on social networks. In: Proceedings of IEEE ICDM 2012, pp. 810–815. IEEE Computer Society (2012)
4. Barabási, A.-L., Albert, R.: Emergence of scaling in random networks. Science **286**(5439), 509–512 (1999)
5. Barbieri, N., Bonchi, E., Manco, G.: Influence-based network-oblivious community detection. In: Proceedings of IEEE ICDM 2013, pp. 955–960. IEEE Computer Society (2013)
6. Batagelj, V., Brandes, U.: Efficient generation of large random networks. Phys. Rev. E **71** (2005)
7. Becker, J., Chen, H.: Measuring privacy risk in online social networks. In: Proceedings of Web 2.0 Security and Privacy (W2SP) (2009)
8. Cavoukian, A.: Privacy by design [leading edge]. IEEE Technol. Soc. Mag. **31**(4), 18–19 (2012)
9. Daley, D.J., Kendall, D.G.: Epidemics and rumours. Nature **208**, 1118 (1964)
10. Erdös, P., Rényi, A.: On random graphs. Publicationes Mathematicae **6**, 290–297 (1959)
11. Erling, O., Averbuch, A., Larriba-Pey, J., Chafi, H., Gubichev, A., Prat-Pérez, A., Pham, M., Boncz, P.A.: The LDBC social network benchmark: interactive workload. In: Proceedings of ACM SIGMOD 2015, pp. 619–630. ACM (2015)
12. Gruhl, D., Liben-Nowell, D., Guha, R.V., Tomkins, A.: Information diffusion through blogspace. SIGKDD Explor. **6**(2), 43–52 (2004)
13. Keeling, M.J., Rohani, P.: Modeling Infectious Diseases in Humans and Animals. Princeton University Press (2008)
14. Kempe, D., Kleinberg, J.M., Tardos, É.: Maximizing the spread of influence through a social network. In: Proceedings of ACM SIGKDD 2003, pp. 137–146. ACM (2003)

15. Kosinski, M., Stillwell, D., Graepel, T.: Private traits and attributes are predictable from digital records of human behavior. PNAS **110**(15), 5802–5805 (2013)
16. Liu, K., Terzi, E.: A framework for computing the privacy scores of users in online social networks. TKDD **5**(1), 6 (2010)
17. Liu, Y., Gummadi, P.K., Krishnamurthy, B., Mislove, A.: Analyzing facebook privacy settings: user expectations vs. reality. In: Proceedings of ACM SIGCOMM IMC '11, pp. 61–70. ACM (2011)
18. Maki, D.P., Thompson, M.: Mathematical models and applications: with emphasis on the social, life, and management sciences. Prentice-Hall (1973)
19. Moreno, Y., Nekovee, M., Pacheco, A.F.: Dynamics of rumor spreading in complex networks. Phys. Rev. E **69**(6), 066130 (2004)
20. Nekovee, M., Moreno, Y., Bianconi, G., Marsili, M.: Theory of rumour spreading in complex social networks. CoRR (2008). arXiv:0807.1458
21. Sabella, R.A., Patchin, J.W., Hinduja, S.: Cyberbullying myths and realities. Comput. Hum. Behav. **29**(6), 2703–2711 (2013)
22. Sudbury, A.: The proportion of the population never hearing a rumour. J. Appl. Probab. 443–446 (1985)
23. Talukder, N., Ouzzani, M., Elmagarmid, A.K., Elmeleegy, H., Yakout, M.: Privometer: privacy protection in social networks. In: Proceedings of M3SN'10, pp. 266–269. IEEE (2010)
24. Wang, Y., Nepali, R.K., Nikolai, J.: Social network privacy measurement and simulation. In: Proceedings of ICNC 2014, pp. 802–806. IEEE (2014)
25. Watts, D.J., Strogatz, S.H.: Collective dynamics of 'small-world' networks. Nature **393**(6684), 409–410 (1998)
26. Wu, L., Majedi, M., Ghazinour, K., Barker, K.: Analysis of social networking privacy policies. In: Proceedings of 2010 EDBT/ICDT Workshops. ACM (2010)
27. Zanette, D.H.: Dynamics of rumor propagation on small-world networks. Phys. Rev. E **65**(4), 041908 (2002)
28. Zheleva, E., Getoor, L.: Privacy in social networks: a survey. In: Social Network Data Analytics, pp. 277–306. Springer, US (2011)
29. Zhou, J., Liu, Z., Li, B.: Influence of network structure on rumor propagation. Phys. Lett. A **368**(6), 458–463 (2007)
30. Zhu, H., Huang, C., Li, H.: Information diffusion model based on privacy setting in online social networking services. Comput. J. **58**(4), 536–548 (2015)

Kosinski, M., Stillwell, D., Graepel, T.: Private traits and attributes are predictable from digital records of human behavior. PNAS 110(15), 5802–5805 (2013)
Liu, K., Terzi, E.: A framework for computing the privacy scores of users in online social [illegible]
[illegible]

Evolution Similarity for Dynamic Link Prediction in Longitudinal Networks

Nazim Choudhury and Shahadat Uddin

Abstract Link prediction problem in network science has spawned not only over myriad applications but also experienced extensive methodological improvements. Different link prediction methods perform feature engineering to build different topological or nodal attribute based metrics measuring the similarity/proximity between non-connected actor pairs to deal with the inference of future associations among them. On the contrary, dynamic link prediction methods have catered the evolutionary process and network dynamics of longitudinal networks. Evolution similarity between node pairs (e.g., similarity between rates of acquiring neighbours by actor pairs over time) can be considered to generate dynamic metrics for the purpose of dynamic link prediction in longitudinal networks. In this study, we attempt to build dynamic similarity metrics by considering the similarity between temporal evolutions of non-connected actor pairs. For this purpose, this study utilises time series forecasting methods to model the temporal evolution of actors' network positions/importance and then it utilizes a dynamic programming method to determine the similarity between these evolutions of actor pairs to quantify the likelihood of future associations among them. Supervised link prediction models exploiting these dynamic similarity metrics were built and performances were compared against some baseline static metrics (i.e., common neighbours). High performance scores achieved by these features, examined in this study, represent them as prospective candidates not only for dynamic link prediction task but also in various applications like security and recommender systems.

N. Choudhury (✉) · S. Uddin
Faculty of Engineering and IT, Centre for Complex Systems Research,
The University of Sydney, Redfern, NSW 2006, Australia
e-mail: Nazim.choudhury@sydney.edu.au

S. Uddin
e-mail: shahadat.uddin@sydney.edu.au

B. Gonçalves et al. (eds.), *Complex Networks VIII*,
Springer Proceedings in Complexity, DOI 10.1007/978-3-319-54241-6_9

1 Introduction

Network science provides various methods supporting the study and modelling of network evolution process that governs network dynamics [1]. Among them, link prediction is the basic and fundamental computational problem that models the underlying growth mechanism of evolving networks [2]. Although link prediction is considered as a time-evolving network analysis model; however, traditional methods generally overlook to take the evolutionary dynamics of the network into account. Abundance of network intrinsic applications and longitudinal network data have triggered the research proliferation in dynamic link predictions. Temporal patterns of longitudinal networks have led scholars to reconsider the evolutionary process in the network over time and utilise these dynamic information in the link prediction task. Recently, researchers have attempted the issue of dynamic link prediction. Temporal link prediction in dynamic networks using time series of topological similarity metrics [3] takes account of the time-aware evolutionary history of topological similarity and employs different forecasting methods. Although different methods of link prediction in temporal networks [4, 5] have generated improved performances in regards to successfully predict future links and/or hidden links, however, some of them are subject to their inherent limitations. For example, probabilistic models involve the prior definition of link occurrences' distribution which is problematic for temporal networks. Further, the exponential random graph model is only suitable for small networks with few hundred nodes. Similarly, matrix or tensor-based methods are not feasible for real-time link predictions in large networks due to the computational complexity and time requirements [6].

Traditional similarity based methods ignored the perception of temporal similarity between actor-based evolutionary information over time for dynamic link predictions. Despite the usage of the temporal pattern of topological similarity metrics, scholars have overlooked the notion of temporal similarity between actor-specific network structural attributes (e.g., network position, network importance) represented by different network measures (e.g., centrality measures). The evolutionary information of these measures can be modelled using time series. Therefore, in this study, we attempt to define a new framework to generate dynamic similarity metrics for link prediction in longitudinal network. These metrics will measure the similarity/proximity between non-connected actor pairs by considering the similarity between their temporal evolutions. Since a longitudinal network can be split into different smaller network snap-shots known as short interval network (SIN), the temporal evolution of an actor can be modelled by using time series of different network measures incident to individual actors in each SIN. The distance between such two temporal sequences, associated with a pair of non-connected actors, is calculated using a dynamic programming method. The resultant distance will define the similarity between the actor pair as like other topological similarity metrics. The research questions we address in this study are: (i) whether the likelihood of future links among non-connected actor pairs depends on their evolution similarity; and (ii) either similar or dissimilar actors, in regards to their evolution, participate in emerging links.

2 Time Series Forecasting Method

In this study, we considered time series of actor-specific network measures to emulate the evolution of actors' positions or behaviours in evolving network. Therefore, we also utilised time series forecasting method to predict their positions in emerging networks in future. In time series forecasting, past observations of a time variable are analysed to develop a model that describes the underlying relationship and extrapolation can be used to predict the future values of the variable. In this study, a well-known forecasting model, known as *ExponenTial Smoothing* (ETS), was used to predict the future centrality measures for every actor under consideration. In this method, forecasts are the weighted averages of previous observations and the weights of the observations decay exponentially with time. Single Exponential Smoothing (SES) with a weight of α is the simplest exponential smoothing method. The forecast equation can be defined as:

$$\hat{y}_t = \alpha y_{t-1} + (1 - \alpha)\, \hat{y}_{t-1}$$

where $\hat{y}_t$ i.e., the forecasted value, depends on both the previous observations and previous forecasts. Linear Exponential Smoothing (LES) refines SES with a β component and considers any short trends in the series. Notably, there are 15 variations of the exponential smoothing process as identified by [7].

3 Dynamic Similarity Metrics

Measuring the degree to which one time series resembles another is a core issue in many mining, retrieval, classification and clustering tasks; however, determining more generic measures reflecting intuitive similarity between pairs of temporal sequences is not straightforward due to its multi-dimensionality [8]. In time series mining, dynamic time warping (DTW) technique [9], a dynamic programming method, is widely used to overcome the limitation of traditional distance measures providing unintuitive distance for two time series where they have approximately the same overall component shapes, but do not align in the time axis.

In this study, a longitudinal network is defined as a time series of network snapshots where each snapshot represents the corresponding network state at a particular time. Each snap-shot is also known as a short interval network. Actors may/may not change their link structures, neighbourhoods and network positions in every SIN over time that can be measured using different network measures utilised in social network analysis [10]. It has been observed that actor based individual network attributes (e.g., network centrality measures) can provide useful information to support link predictions [11]. Therefore, in this study, we utilised time series of *degree centrality* and .itcloseness centrality measures to emulate an actor's network dynamicity. Further extension of this study can be achieved by introducing other actor-based network measures (e.g., betweenness centrality).

Considering total N number of SINs, let X_u and Y_v be the time series of length $|m|$ and $|n|$ considering network measure c for actor u and v of a link (u, v), where $m, n \leq N$. Thus, two time series associated with actor u and v can be written as:

$$X_u(c) = x_1, x_2, x_3, x_4 \ldots\ldots\ldots x_m, \qquad Y_v(c) = y_1, y_2, y_3, y_4 \ldots\ldots\ldots y_n$$

Here x_i and y_i denote the value of c for actor u and v in SIN G_t where $t = 1, 2, 3 \ldots\ldots N$. A local cost/distance measure $d(x_i, y_i)$, usually from one of the traditional well-defined distance measures (e.g., Euclidean, Minkowski), is defined to compare two different points in X_u and X_v. Typically, the cost/distance provided by this measure is small if the corresponding point x_i in X_u and y_i in Y_v is similar to each other and large otherwise.

Our goal here is to find an alignment between X_u and Y_v, having minimal overall cost. The notion of this alignment depends on the definition of an (m, n)-warping path which is a sequence $p = p_1, p_2, p_3, \ldots p_l$ with $p_l = (m_l, n_l) \in [1 : m] \times [1 : n]$ for $l \in [1 : L]$. The optimal warping path between X_u and Y_v is defined as a warping path p^* with the minimal cost among all possible warping paths. Therefore, the value of our dynamic similarity metric for node pair u and v is defined as:

$$d_{p^*}(x_i, y_j) = \min\{\sum_{l=1}^{L} d(x_{ml}, y_{nl})\,|p \text{ is an } (m, n) \text{ warping path}\}$$

In Fig. 1, we depict the visual representation of our framework to generate dynamic similarity metrics. In this figure, the solid green and red lines represent the network measures at each SIN during training period and the dotted lines represent the forecasted network measure. The black arrowed lines represent the mapping path utilised to measure the similarity between actor u and v using dynamic programming method. The final proximity or similarity score calculated by dynamic similarity metrics is generated by the accumulated distance cost of this optimal mapping path.

4 Longitudinal Datasets

For longitudinal network data, this study considers three datasets from the 'Network Repository' [12] which is the first and the largest interactive repository of network dataset. From the dynamic dataset category, we extracted three datasets tagged as 'manufacturing-email', 'facebook-messages' and 'facebook friendship graph' where links between node pairs are time-stamped. The first dataset contains internal email communications among employees of a mid-sized manufacturing company. The second dataset contains network data from a Facebook-like social network originated from an online community for students at University of California, Irvine,

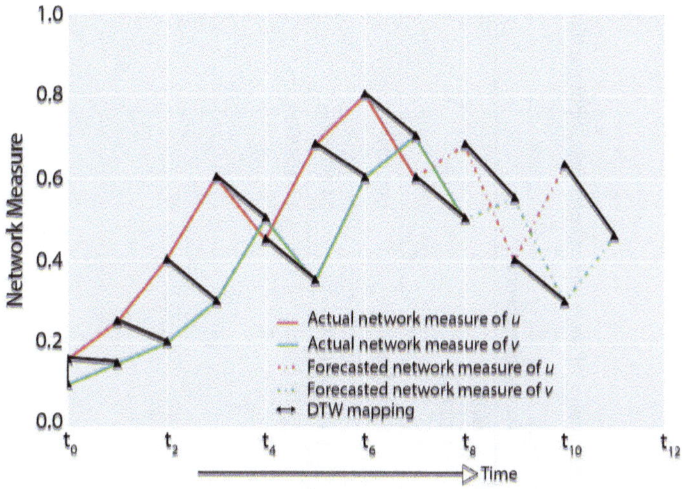

Fig. 1 A visual representation of the framework to generate dynamic similarity metrics. The *solid green* and *red lines* represent network measures (e.g., degree centrality) of node u and v in short interval networks during the training phase. The *dotted lines* represent the forecasted network measures during the test phase. The *black lines* represent the mapping path considering similar points of two time series using dynamic time warping technique

where actors represent students within the community and links represent messages communicated among them. The final dataset is comprised of a real world Facebook friendship network in which Facebook users are actors and friendship relations among users are links.

For the sake of brevity, in the rest of the study, we name our three datasets as 'Email Network', 'UCI Network' and 'Facebook Network' for three datasets respectively. For the purpose of the supervised link prediction, the range of temporal networks was divided into two non-overlapping sub-ranges; i.e., the training phase and test phase. We also split our training network dataset into smaller temporal graphs considering time window size of 1 day, 2 days, 7 days and 30 days to generate our SINs. Table 1 describes the basic statistics of our datasets including total number of actors and links, the duration of the training and test phases in dates and the number of SINs in the training phase using four different time granularity (e.g., 30 days).

5 Supervised Link Prediction

Supervised methods for link prediction problems can predict possible future links by successfully discriminating between the links with positive and negative labels within a classification dataset. For modelling supervised link prediction, we followed the model described in [11] including the workload ratio of links with positive and negative labels to 1:10. Loops and duplicate links were ignored. In this study, we

Table 1 Basic statistics of longitudinal network datasets utilised in this study. The term SIN represents the number of short interval networks. Split duration denotes the time interval used to split the training network into network snap-shots

	Email network				UCI network				Facebook network			
Nodes	167				1899				6988			
Total edges	3251				15737				14736			
Training duration	Start date		End date		Start date		End date		Start date		End date	
	02-01-2010		03-09-2010		15-04-2004		28-09-2004		14-10-2004		24-08-2006	
Test duration	04-09-2010		30-09-2010		29-09-2004		26-10-2004		25-08-2006		14-10-2006	
Split duration	1	2	7	30	1	2	7	30	1	2	7	30
No. of SINs	215	121	35	9	165	83	24	7	597	322	97	23

utilised dynamic similarity metrics considering network dynamics, as described in Sect. 5, to describe instances in classification datasets. For the validation purpose and comparisons sake, we utilised two baseline static metrics ignoring the network dynamics and considering only a static network constructed by the aggregation of all temporal networks within the training phase. These two baseline static features are Common Neighbours [13] and AdamicAdar [14]. The feature values were normalized with zero mean and one standard deviation. This study also used simple logistic regression, Naïve Bayes, and Random Forest algorithms for classification purposes. Performances of these classifiers were then compared using different performance metrics including a 10-fold cross-validation and the mean score as accuracy percentage, AUCROC (Area under Receiver Operating Characteristics Curve) and AUCPR (Area under Precision-Recall Curve).

6 Results

In Table 2, we present the performance displayed by our four classifiers in classifying edges with positive and negative labels using both dynamic and static features. In this table, we considered the short interval networks generated using split duration of seven days (see Table 1) during the training period. In regards to three performance measurement metrics, we found that our dynamic metrics performed as good as the baseline metrics and in some cases outweighed them (e.g., UCI network). Despite their underperformances in regards to AUCROC in the Facebook dataset, scores are better in comparison to a random algorithm having highest AUCROC score

Table 2 Performance measurements of different classifiers using dynamic and static features. Different classifiers are LR = Logistic Regression, NB = Naive Bayes and RF = Random Forest

	Dynamic metrics				Static metrics		
	Classifier	Accuracy (%)	AUC ROC	AUC PR	Accuracy (%)	AUC ROC	AUC PR
Email network	LR	86.3	0.87	0.43	91.2	0.92	0.74
	NB	86.2	0.83	0.39	89.9	0.92	0.73
	RF	87.2	0.89	0.52	92.3	0.94	0.79
UCI network	LR	91.8	0.83	0.39	91.4	0.60	0.21
	NB	91.3	0.77	0.36	90.7	0.60	0.21
	RF	92.0	0.84	0.43	91.3	0.61	0.19
Facebook network	LR	91.2	0.60	0.12	92.2	0.66	0.51
	NB	91.2	0.56	0.12	91.8	0.66	0.50
	RF	91.3	0.66	0.16	93.1	0.66	0.52

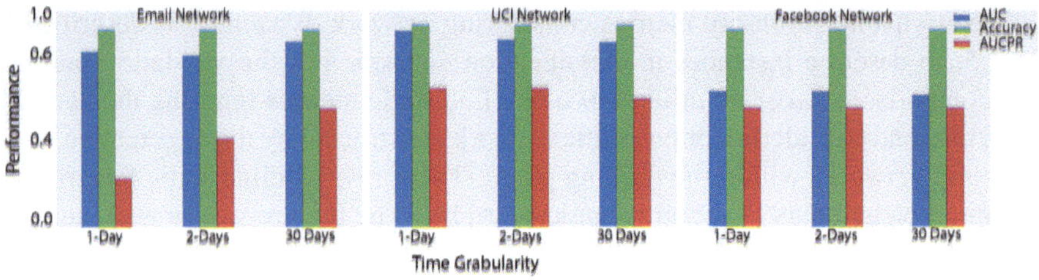

Fig. 2 Performance measures by Random Forest classifier considering short interval networks of different time granularity in three network datasets

of 0.50. Notably, in this dataset, low AUCROC scores were observed using static metrics too. The authors in [15] proposed a method to determine the minimum value of AUCPR as $AUCPR_{min} = 1 + \frac{(1-\pi)\ln(1-\pi)}{\pi}$ with skew $\pi = \frac{positive\ samples}{n}$ where n = total number of samples in the classification dataset. According to this equation, with skew $\pi = 0.091$ (since the ratio of positive and negative samples is $1:10$ in this study), the minimum value of AUCPR should be 0.04. In Table 2, we also observed that most of the classifiers outweigh this minimum value utilising our dynamic features. In Fig. 2, we represent the performance scores by Random Forest classifier considering network snap-shots of different time granularity (i.e., one day, two days and 30 days). We also observe high performance scores across three dataset. These results answer our first research question, described in the introduction section, that temporal evolution similarity between actor pairs can be a potential candidate for link prediction in dynamic networks.

In order to answer our second research question, we analysed the distance between the time series of network measures, incident to non-connected actor pairs of true links in the test phase, calculated by the DTW method in this study. Alternatively, the values of our dynamic feature for the positively labelled links in the classification datasets were examined. In Fig. 3, we visualize the distribution of dynamic feature values of the actual links in the test phase. From this figure, it is observable that actors with minimum distance between their evolutionary measures participated in the actual links those occurred in the test phase. Alternatively, similar actors, in regards to their dynamics, have higher likelihood in forming emergent links.

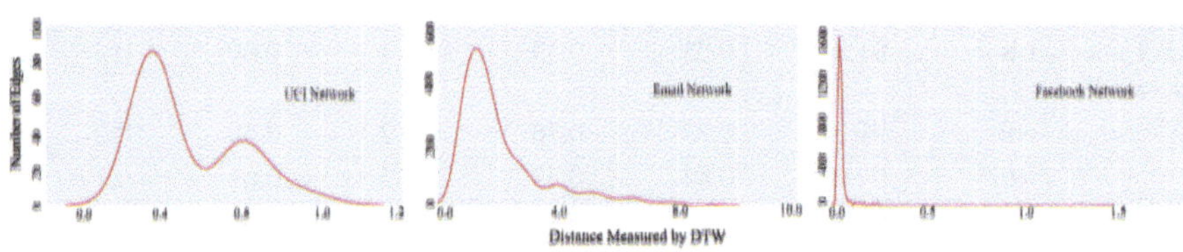

Fig. 3 Distance measured by DTW method between the temporal evolutions of true links in the test phases of three datasets

7 Discussion and Conclusion

In this study, we have observed that dynamic network topology along with associated evolutionary information resulting from the temporal and structural changes of individual actors, can be exploited in dynamic link predictions. Further, since most networks inherently evolve over time, scholars delved into temporal networks and network dynamics to resolve issues with link prediction problem in dynamic networks [16]. In this regard, we demonstrated in the result section that our dynamic features perform as high as the static features those are widely used in traditional link predictions. In some cases, they outweighed the static features to predict future links in longitudinal networks. We validated our assumption and results of link prediction by considering well-known performance metrics utilised in supervised link prediction. Our empirical analysis found that actors, with similarity in their dynamicity, have the most likelihood of forming links in future. This study can further be extended considering various aspects. For example, different other network measures and well-known forecasting methods like ARIMA can be exploited to boost the performance of link prediction. Considering high performances, as described in the result section, the dynamic features, built in this study, can be applied to model security network or network-based recommender system. For example, the buying behaviour of consumers can be explored over time to predict the future associated purchases. Measuring the temporal similarity of associated buying patterns may help determining new patterns of associated purchases.

References

1. Opsahl, T., Hogan, B.: Growth mechanisms in continuously-observed networks: Communication in a facebook-like community. arXiv:10102141 (2011)
2. Liben-Nowell, D., Kleinberg, J.: The link-prediction problem for social networks. J. Am. Soc. Inf. Sci. Technol. **58**(7), 1019–1031 (2007)
3. Soares PRdS, Prudêncio RBC Time series based link prediction. In: The 2012 International Joint Conference on Neural Networks (IJCNN), pp. 1–7. IEEE (2012)
4. Pujari, M., Kanawati, R.: Supervised rank aggregation approach for link prediction in complex networks. In: Proceedings of the 21st International Conference on World Wide Web, pp 1189–1196. ACM (2012)
5. Hanneke, S., Fu, W., Xing, E.P.: Discrete temporal models of social networks. Electron. J. Stat. **4**, 585–605 (2010)
6. Ibrahim, N.M.A., Chen, L.: Link prediction in dynamic social networks by integrating different types of information. Appl. Intell. **42**(4), 738–750 (2015)
7. Hyndman, R.J., Athanasopoulos, G.: Forecasting: principles and practice. OTexts (2014)
8. Müller, M.: Dynamic Time Warping. Information Retrieval for Music and Motion. Springer, Berlin Heidelberg (2007)
9. Vintsyuk, T.K.: Speech discrimination by dynamic programming. Cybern. Syst. Anal. **4**(1), 52–57 (1968)
10. Uddin, S., Khan, A., Piraveenan, M.: A set of measures to quantify the dynamicity of longitudinal social networks. Complexity **21**(6), 309–320 (2016)
11. Choudhury, N., Uddin, S.: Time-aware link prediction to explore network effects on temporal knowledge evolution. Scientometrics **108**(2), 745–776 (2016)

12. Rossi, R.A., Ahmed, N.K.: Networkrepository: a graph data repository with visual interactive analytics. In: 29th AAAI Conference on Artificial Intelligence, Austin, Texas, USA, 25-30 January 2015. Association for the Advancement of Artificial Intelligence, pp. 4292–4293
13. Newman, M.: Clustering and preferential attachment in growing networks. Phys. Rev. E **64**(2), 25102 (2001)
14. Adamic, L.A., Adar, E.: Friends and neighbors on the Web. Soc. Netw. **3**(25), 211–230 (2003)
15. Boyd, K., Costa, V.S., Davis, J., Page, C.D.: Unachievable region in precision-recall space and its effect on empirical evaluation. In: Machine Learning: Proceedings of the International Conference. International Conference on Machine Learning, p. 349. NIH Public Access (2012)
16. Xu, H.H., Zhang, L.J.: Application of link prediction in temporal networks. In: Advanced Materials Research, pp 2231–2236. Trans Tech Publication (2013)

Stochastic Modeling of the Decay Dynamics of Online Social Networks

Mohammed Abufouda and Katharina A. Zweig

Abstract The dynamics of online social networks (OSNs) involves a complicated mixture of growth and decay. In the last decade, many online social networks, like MySpace and Orkut, suffered from decay until they were too small to sustain themselves. Thus, understanding this decay process is crucial for many scenarios that include: (1) Engineering a resilient network, (2) Accelerating the disruption of malicious network structures, and (3) Predicting users leave dynamics. In this work we are interested in modeling and understanding the decay dynamics in OSNs to handle the aforementioned three scenarios. Here, we present a probabilistic model that captures the dynamics of the social decay due to the inactivity of the members in a social network. The model is proved to have *submodularity* property. We provide preliminary results and analyse some properties of real networks under decay process and compare it to the model's results. The results show, at the macro level of the networks, that there is a match between the properties of the decaying real networks and the model.

1 Introduction

Today's online social networks represent a main source of communication and information exchange among people all over the world. Many online social networks have proven their usefulness, like Facebook, Twitter, and Linkedin, in connecting people and facilitating an exquisite new medium for sharing news, forming groups of people of the same interests, and eliciting knowledge. The growth of these networks in terms of user activity shows that these online social networks have become a vital part in today's human activities. One well studied aspect of online social networks dynamics is the *growth* phenomenon of a network. The work by Barabási

M. Abufouda (✉) · K.A. Zweig
Computer Science Department, University of Kaiserslautern,
Gottlieb-Daimler-Str. 48, 67663 Kaiserslautern, Germany
e-mail: abufouda@cs.uni-kl.de

K.A. Zweig
e-mail: zweig@cs.uni-kl.de

B. Gonçalves et al. (eds.), *Complex Networks VIII*,
Springer Proceedings in Complexity, DOI 10.1007/978-3-319-54241-6_10

et al. [6] presented a simple model for understanding the growth dynamics of a network, namely the *Preferential Attachment Model* (PAM), which is a rich-get-richer-model. Jin et al. [15] noticed that the model by Barabási et al. [6] and other similar models, like the work by Dorogovtsev et al. [11] for modeling the growth of random networks, are not suitable to understand the growth dynamics of social networks. Thus, they provided a model that considers the specialty of social networks without a power law distribution and with large clustering coefficient [15]. With the availability of the online datasets, Newman [25] studied empirically the growth of social networks using the scientific collaboration networks against the PAM model [6]. Bala et al. [5] provided a non-cooperative game based model for the network formation. Later, Jackson [14] surveyed the models and methods that were used to capture the network formation process and compared them in terms of stability and efficiency. Leskovec et al. [21] first showed on dynamic network data, that networks densify over time and that their diameter is shrinking. They also provided another growth dynamics model that was able to produce networks with these properties. The previous work and the availability of rich datasets pushed the research to an in-depth investigation of the properties of the networks over time. Kumar et al. [20] studied the growth of a large social network in terms of network component analysis, Kossinets et al. [18] studied the tie formation process within the social networks that is affected by internal and external factors, and Capocci et al. [9] studied the statistical properties of the growth characteristics of Wikipedia collaboration social networks. Likewise, Backstrom et al. [4] studied empirically how groups are formed and evolve over time in MySpace social networks and Mislove et al. [23] provided a study for the growth of Flicker social network. Even though there are many successful social networks, the evolution of a social network also incorporates decay. In the last decade, some of the online social networks were closed after a huge loss or inactivity of their members. Online social networks, like Friendsfeed, Friendster, MySpace, Orkut, and many websites of the Stack Exchange platform, are now out of service, despite the fact that some of them, e.g., Orkut and Myspace, showed a tremendous growth [2] just a decade ago. The decay of these networks poses many questions about the reasons behind their fall down. Garcia et al. [12] and Chhabra et al. [10] studied the static properties of Friendster and MySpace, respectively, in order to understand the network-related properties of these networks as an example of a decayed network. Recent studies by Malliaros et al. [22] and Bhawalkar et al. [8] provided theoretical models for understanding the social engagement in online social networks with a potential to predict social inactivity. Torkjazi et al. [28] provided an experimental analysis of Myspace online social network and examined the activity and inactivity of its users with some insights about the reasons behind the fall of MySpace. Similarly, Ribeiro [26] studied activity and inactivity of the users by providing a model that uses the number of daily active users as a proxy of the dynamics in the membership based websites. Kairam et al. [16] provided machine learning prediction models to predict community *longevity*: how long a community in an online social network will survive. Another related work done by Asur et al. [3] discussed the persistence and decay of Twitter tweets. While investigating the reasons behind the inactivity of members of an online social networks is not in the scope of this work, some recent studies proposed some answers [17, 27], suggesting that the main reason behind this decay is the inactivity of the members of the online social networks.

Building a sound understanding of the decay dynamics of networks requires not only studying the static properties of these networks, but also requires investigating their dynamics and properties over time, and this is what we are interested in here. We consider the Stack Exchange websites that were closed after some period of time due to the lack of enough activity required to keep the website alive. The closed websites are an example of the social network decay, where we model the members of a website as the nodes of the network and an edge exists between any two nodes if they post, comment, or answer to the same question in the websites.

While we cannot answer why a person starts losing interest in a social network, we can try to analyze and model the effect of this behavior on other people. Such a model might in turn hint at the causes of social decay or at least explain some part of it.

In this work, we provide a probabilistic model for understanding the social decay phenomenon in online social networks. The model presented here can provide insights regarding the effect of node leave on the neighborhood nodes. Our contribution in this work is split the following: (1) A longitudinal network analysis of the stack exchange sites showing their decay. (2) A probabilistic model for social network decay which is a *step by step* mechanistic model for a node leave and the effect of its leave. (3) Theoretical proof of the submodularity of the model that leads to viable optimization, e.g., determining the minimal set of nodes to leave the network for accelerating/decelerating the decay process. Being submodular renders the maximization problem of the model to be viable.

2 Model and Notations

A network $G = (V, E)$ is a tuple of two sets V and E, where V is the set of nodes and E is the set of edges such that an undirected edge e is defined as $e = \{u, v\} \in E$, where $u, v \in V$. As we consider a dynamic system, the notation G^t is a network at time t. We assume that every node $w \in V$ has an initial *Leave Probability* $\pi_w^{t=0}$ which denotes the probability of node w leaving the network at time 1, and generally at $t+1$. If a node w did not leave at $t+1$, i.e., $w \in V(G^{t+1})$, then its current leave probability, π_w^t, will be increased depending on its neighbors who left at $t-1$. The *tie strength* at time $t-1$, representing some possibly dynamic measure of closeness of a relationship, is denoted by $\delta_{v,w}^{t-1}$ and assumed to be $\in (0, 1]$. The details of this process are described in the following sections.

Definition 1 A dynamic network G is called a "Decaying Network" if $|E(G)^{t-1}| \geq |E(G)^t|$, $|V(G)^{t-1}| \geq |V(G)^t|$, and $V(G)^t \subseteq V(G)^{t-1}$, $\forall t > 0$.

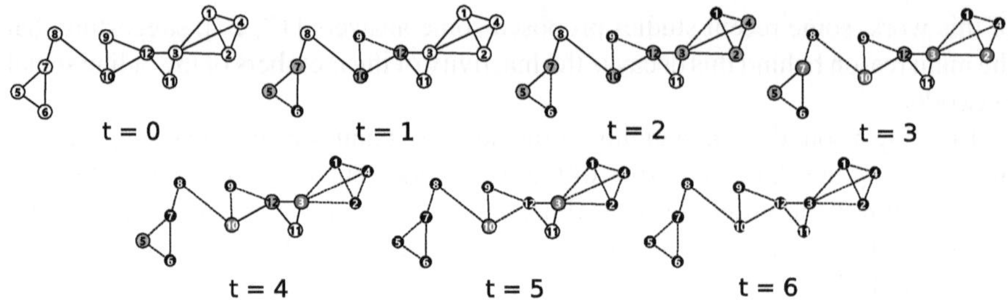

Fig. 1 An illustration of the model. The color of the nodes represents how likely a node will leave in the future, where *white* nodes are very unlikely to leave and the level of grayness correlates with the probability to leave. Whenever a node leaves the network it is marked as *black*, all its edges are removed, and all of its neighbors get affected by its leave by increasing their leave probability. The *dotted* edges are the removed edges

We assume the model starts with a *Decaying Network*, i.e., no further nodes or edges are added to the network. The main idea of the model is shown in Fig. 1.

2.1 Probability Gain

At any point of time t where $t > 0$, the node's leave probability changes from π_w^{t-1} to π_w^t, by adding *Probability Gain* $\Delta\pi_w^t$, that never exceeds the value of 1. Thus, a node w will leave at time $t+1$ with probability π_w^{t+1} such that:

$$\pi_w^{t+1} = min\{1,\ \pi_w^{t-1} + \Delta\pi_w^t\} \tag{1}$$

If a node w did not leave the network at time t, then we have two sets: $\overline{\Gamma}_w^{t-1}$ and $\underline{\Gamma}_w^{t-1}$, which are the sets of w's neighbors who left and did not leave the network at $t-1$, respectively.

2.1.1 Probability Gain Due to One Node Leave

We first define the probability gain due to the leave of a single neighbor v of the node w at time point $t-1$, and then generalize it to w's neighbors that left the network: $\overline{\Gamma}_w^{t-1}$. Now, the probability gain that a node w will get at $t+1$ due to the leave of its neighbor node v at $t-1$ is defined as:

$$\Delta\pi_w^{t+1}(v) = 1 - (1 - \pi_v^{t-1})(1 - \delta_{v,w}^{t-1}) \tag{2}$$

where the edge $e = (v, w) \in E(G)_{t-2}$ and $e = (v, w) \notin E(G)_{t-1}$ as $v \in \overline{\Gamma}_w^{t-1}$ and $w \in V(G)_{t-1}$. Thus, the total probability gain produced by the leave of node v to all

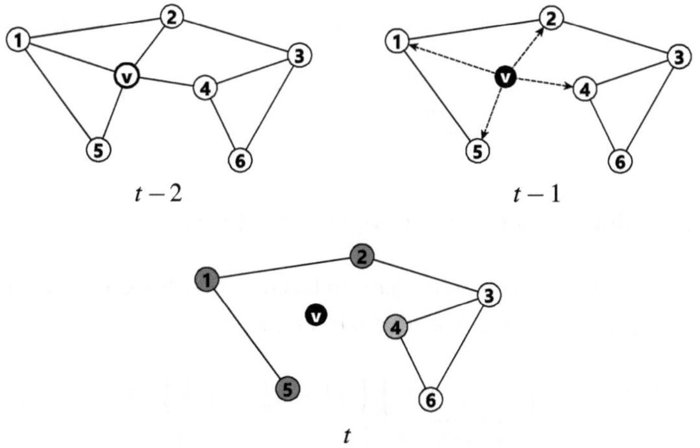

Fig. 2 This figure shows how a node v affects all of its neighbors when it leaves. At $t-2$, the node v has a leave probability π_v^{t-2} which was gained by v's initial leave probability π_v^0 and possible probability gains caused earlier by leaving neighbors, i.e., $\pi_v^{t-2} = \pi_v^0 + \sum_{t=1}^{t=t-3} \Delta\pi_v^t$. At time $t-1$, the node v leaves the network affecting its neighbors by increasing the leave probability of nodes 1, 2, 4, 5. Here we assume that the tie strength between v and the nodes 1, 2, 5 is greater than the tie strength between v and 4. That is why the nodes 1, 2, 5 gain more leave probability than node 4, which is represented by a *darker* color of nodes 1, 2, 5

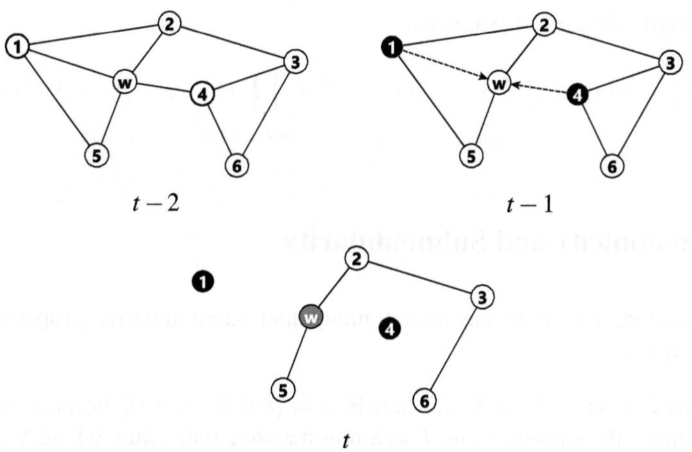

Fig. 3 This figure shows how a node w is affected by the leave of its neighbors. At $t-2$, the nodes 1, 4 have leave probabilities π_1^{t-2} and π_4^{t-2}, respectively, which were gained by the nodes' initial leave probabilities π_1^0 and π_4^0 and possible earlier probability gains. At time $t-1$, the nodes 1, 4 leaves the network affecting their neighbors, here we are interested in the node w. The leave of nodes 1, 4 left node w with an increased leave probability at t. Note that nodes 2, 3, 5, 6 are affected also by the leave of 1, 4, but for simplicity and for visualization traceability we concentrated on node w

of its neighbors which did not leave, see Fig. 2 for an illustration, is given by (Fig. 3):

$$\Delta\pi^t(v) = \sum_{w \in \underline{\Gamma}_v^{t-1}} 1 - (1 - \pi_v^{t-1})(1 - \delta_{v,w}^{t-1}) \tag{3}$$

2.1.2 Probability Gain Due to Multiple Nodes Leave

We now generalize the probability gain induced by the leave of a single node to capture the impact of all neighbors that left, i.e., $\overline{\Gamma}_w^{t-1}$.

$$\begin{aligned}\Delta\pi_w^t &= 1 - [\underbrace{(1 - \xi_w^{t-1})}_{\text{Assures leave}}(\underbrace{\prod_{u \in \overline{\Gamma}_w^{t-1}} (1 - \pi_u^{t-1})}_{\text{Leave probabilities effect}})(\underbrace{\prod_{u \in \overline{\Gamma}_w^{t-1}} (1 - \delta_{u,w}^{t-1})}_{\text{Tie strength effect}})] \\ &= 1 - [(1 - \xi_w^{t-1})(\prod_{u \in \overline{\Gamma}_w^{t-1}} (1 - \pi_u^{t-1})(1 - \delta_{u,w}^{t-1}))]\end{aligned} \tag{4}$$

where $\xi_w^{t-1} = \frac{|\overline{\Gamma}_w^{t-1}|}{|\Gamma_w^{t-1}|}$ and the quantity $1 - \xi_w^{t-1}$ assures that when all of the neighbors of the node w leaves, then the node w will (be forced to) leave too as it will be disconnected. Thus, Eq. 1 becomes:

$$\pi_w^t = min\{1,\ \pi_w^{t-1} + 1 - [(1 - \xi_w^{t-1})(\prod_{u \in \overline{\Gamma}_w^{t-1}} (1 - \pi_u^{t-1})(1 - \delta_{u,w}^{t-1}))]\} \tag{5}$$

3 Monotonicity and Submodularity

In this section, we show the monotonicity and submodularity properties of the model's equations.[1]

Definition 2 Let $f : 2^V \rightarrow \mathbb{R}_{\geq 0}$, where $\mathbb{R}_{\geq 0} = \{x \in \mathbb{R} \mid x \geq 0\}$, be an arbitrary function that maps the subsets S and T to a non-negative real value, where $S \subseteq T \subset V$. Then, the function f is submodular [19] if it satisfies the following inequality: $f(S \cup \{v\}) - f(S) \geq f(T \cup \{v\}) - f(T)$, where $v \in V \setminus T$.

Lemma 1 (Order preserving of the probability gain sum) *Let $\pi^t = \{\pi_1, \pi_2, \cdots, \pi_n\}$, where $\pi_i \in \pi^t$ and $\pi_i \in (0, 1]$. Then we have: $\sum_{\pi_i \in \pi^t} \pi_i \leq \sum_{\pi_i \in \pi^{t+1}} \pi_i$ where $\pi^t \subseteq \pi^{t+1}$, and the sets π^t and π^{t+1} are defined like above.*

[1] Detailed proofs are provided in an earlier technical paper [1].

Lemma 2 (Order preserving of the probability gain product) *Let* $\pi^t = \{\pi_1, \pi_2, \cdots, \pi_n\}$, *where* $\pi_i \in \pi^t$ *and* $\pi_i \in (0, 1]$. *Then we have:* $\prod_{\pi_i \in \pi^t} \pi_i \geq \prod_{\pi_i \in \pi^{t+1}} \pi_i$ *where* $\pi^t \subseteq \pi^{t+1}$, *and the sets* π^t *and* π^{t+1} *are defined like above.*

Theorem 1 *The leave probability gain function, Eq. 3, is submodular.*

The interpretation of the theorem is that, the more friends a node v had before leaving, the higher its total induced leave probability gain.

Theorem 2 *The leave probability gain function, Eq. 4, is monotone, i.e., for a node* w *we have* $\pi_w^t \leq \pi_w^{t+1}$ *if the node* w *did not leave the network at* $t + 1$.

Theorem 3 *The leave probability gain function, Eq. 4, is submodular.*

The theorem state that the more of your friends leave, the less important the others become. Submodulariy entails an interesting properties: the minimization problem of submodular function can be performed in polynomial time [13], and the maximization problem of the submodular function, which is NP-Hard problem, can be approximated within a factor of $\alpha = (1 - 1/e)$ using a greedy algorithm [24].

4 Results

In this section, we provide the analysis of the decaying stack exchange websites and the results of the model. Figure 4a shows the distribution of the number of user comments for alive and decayed websites. The figure shows that the decayed websites clearly have different distribution characteristics with a low mean and low standard deviation. A similar behavior is found in Fig. 4b, c that represents the distribution of users' total received *Reputation* and *Upvotes*, respectively. These two properties reflect the level of knowledge and experience that the members of a website have. For the decayed websites, it is clear that, on average, the members have much less reputation and upvotes than those in the alive websites. The three figures, Fig. 4a–c show that there is less social activity in the decayed websites, which may be used as an indication for studying the future of the alive websites. However, understanding the decay dynamics of the decayed websites requires a deeper investigation and modeling for the nature of the interaction among the members. Our approach to better understand what happens during the decay process is to make a network representation of the members' interactions, like comments, upvotes, and posts, as networks. Then, we build a network based model for modeling the decay process. Algorithm 2 depicts the steps we followed in our experiments. Line 4 initializes the initial leave probability π_v^0, which is a design decision and we selected values from 0.0005 to 0.045 with an 0.0005 increase step. For each of these values, the model runs and simulates Eq. 4. The update step in line 13 simulates Eq. 5. The result of the algorithm is a set of graphs that are used for the analysis. The output of this algorithm results in a large number of graphs. For example in the case of the Startup Business website

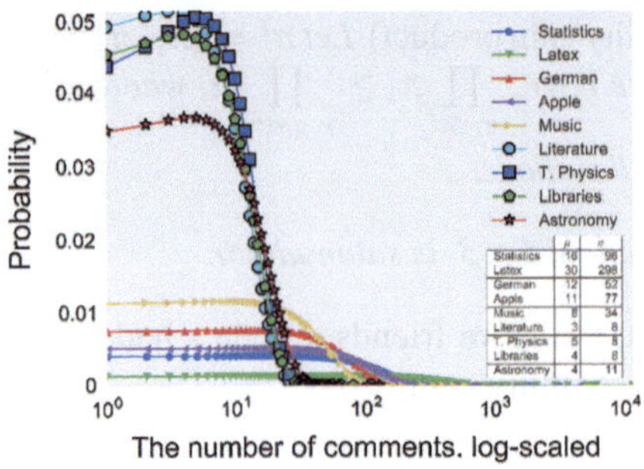

(a) Comments per user.

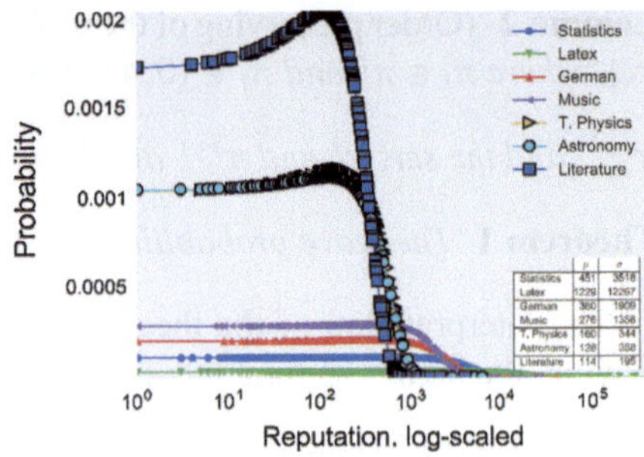

(b) Reputation per user.

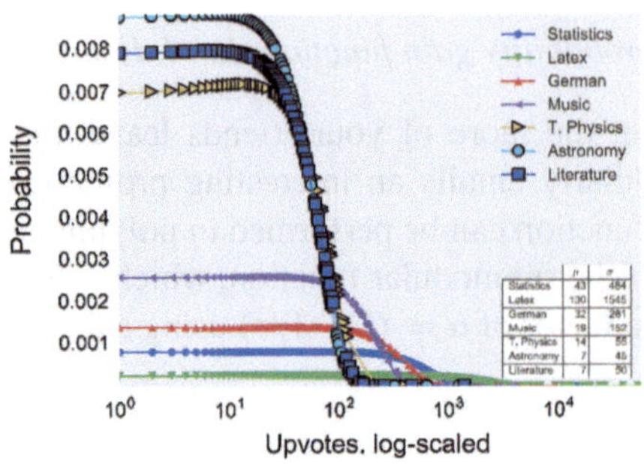

(c) Upvotes per user.

Fig. 4 The characteristics of the interaction decay in the decayed and alive websites of the Stack Exchange websites. The figures show the probability distributions of different types of interactions in these websites. Markers with *bold boarders* are decayed websites, μ is the mean, and σ is the standard deviation. From the figures it is clear that the decayed networks have different distribution properties from the other alive networks

we have analyzed more than 200*k* graphs with 250 runs for each probability to get more confidence of the results. The tie strength was a normalized edge weight where the weight is the frequency of the interaction between two nodes.

In Fig. 5 we show the macro properties of the real networks of the Startup Business website over time. The network evolution shows a clear decay that is represented as a decrease in the number of the nodes. This decrease was associated with a decrease in the average degrees of the nodes over time and also with a decrease of the node's coreness [7]. Another macro measure we used is the network density. Figure 5c shows an increase in the density over time. This increase is due to early leave of the nodes with less degrees, i.e., the nodes that are part of dense subgraphs seem to leave the network late. Now, we will show the results of the model simulation. Figure 6a shows the number of components in the network over simulation for different values of π_v^0. The number of components start to increase to a maximum value before it start to decrease. The reason is that at the beginning the model starts with a one-connected component graph and after each step some nodes are removed due to the leave probability. The leave of some nodes results in a disconnected graph with more

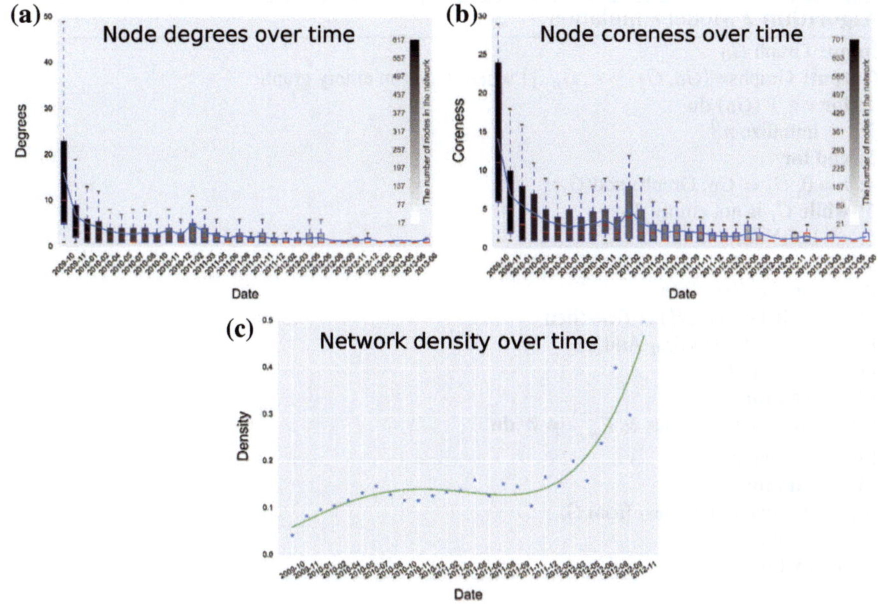

Fig. 5 Macro properties of the real networks under decay for the Startup business site. **a–c** Show the degrees of the nodes, the node coreness, and the network density over time

components. The number of these disconnected components increases until these disconnected components are composed of only triples or simple edges. As a result, a node that leaves from these triples or from these edges will not increase the number of the components anymore. Figure 6b, c show a similar behaviour for the average degree and the average coreness over time, respectively. The more nodes are being removed from the network, the less edges remain and thus the average degree and the average coreness decrease uniformly over time. This behavior of the model is similar to the real data presented in Fig. 5. The last global measure that we use is the network density as shown in Fig. 6d. The density of the simulated networks increases over time for the same reason stated for the real networks in Fig. 5. These results show that the model provides a real-like behaviour of the networks under decay.

Algorithm 2 Model simulation

Input: Graph G_0
Output: Graphs= $\{G_0, G_1, \cdots, G_{n-1}\}$ where G_n is an empty graph
for $v \in V(G_0)$ **do**
  initialize π_v^0
end for
$t = 0$, $G_t = G_0$, Graphs.add(G_t)
while G_t is not empty **do**
  LeftNodes$_t = \emptyset$
  $t = t + 1$
  for $v \in V(G_t)$ **do**
    if Leave(v,π_v^t) is *True* **then**
      LeftNodes$_t$.Add(v)
    end if
  end for
  for $u \notin$ LeftNodes & $\overline{\Gamma}_u^{t-1} \neq \emptyset$ **do**
    update(π_u^t,$\overline{\Gamma}_u^{t-1}$)
  end for
  remove LeftNodes$_t$ from G_t
  Graphs.add(G_t)
end while

5 Discussion

There are different applications where the model can be utilized. *1. Social network resilience*: the resilience against huge disruptions in social networks is not well-studied. We think that the model provides a first step towards engineering a resilient social network via understanding the decay dynamics of a network. *2. Leave cascade detection*: the leave of one member is not as harmful as a cascade of leaves for the networks that seek growth. The model captures the dynamics of leave cascades by observing the leave probabilities of the nodes and their increase. *3. Maximizing the leave effect*: for a network where a dissolving process is required, like criminal social networks, the model is able to provide a viable disruption maximization (thanks to the submodularity property of the model) to the network with insights about the influential members and the effect of the leave.

6 Conclusion

In this work, we presented an empirical analysis of the social decay dynamics of the closed Stack Exchange websites. The closed websites showed an inactivity, which might have caused their decay. We model these interactions between the members of these websites as a network that enabled us to build a model to understand the decay dynamics. Then, we have presented a model for capturing the decay dynamics

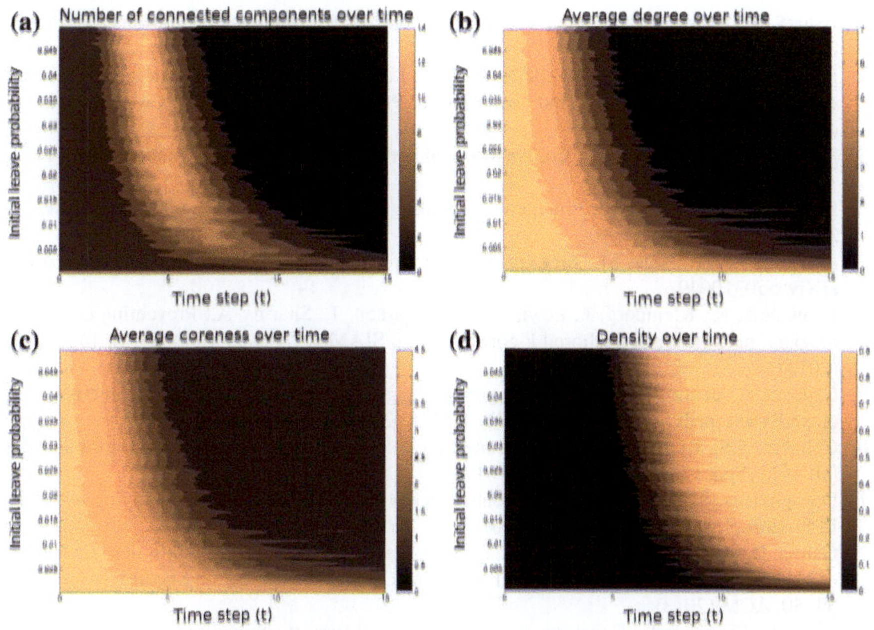

Fig. 6 The results of multiple global measures of the model. **a–d** Show the number of components, the average degree, the average coreness, and the density of the network over time for different values of initial leave probability π_v^0, respectively. The model started with G_0 as the input network and simulates the decay over it

in social networks. The model is a probabilistic model that assumes that the leave of social network members affects the leave of their neighbors. In this work we have also presented some mathematical properties and proved them. We proved that the model's main equations are submodular, which entails doing optimization of the model in a feasible way. Also, we presented the macro network properties of real networks under decay and compared these results with the results of the model simulation. The results of the model and the real networks under decay showed a similar behavior that supports the potential of the model for different usages. In the future, we will design the optimization algorithms and study the applicability of the model and also provide more empirical validation of its properties.

References

1. Abufouda, M., Zweig, K.A.: A theoretical model for understanding the dynamics of online social networks decay (2016). arXiv:1610.01538
2. Ahn, Y.-Y., Han, S., Kwak, H., Moon, S., Jeong, H.: Analysis of topological characteristics of huge online social networking services. In: Proceedings of the 16th International Conference on WWW, pp. 835–844. ACM (2007)

3. Asur, S., Huberman, B.A., Szabo, G., Wang, C.: Trends in Social Media: Persistence and Decay (2011). SSRN 1755748
4. Backstrom, L., Huttenlocher, D., Kleinberg, J., Lan, X.: Group formation in large social networks: membership, growth, and evolution. In: Proceedings of the 12th ACM SIGKDD, pp. 44–54. ACM (2006)
5. Bala, V., Goyal, S.: A noncooperative model of network formation. Econometrica **68**(5), 1181–1229 (2000)
6. Barabási, A.-L., Albert, R.: Emergence of scaling in random networks. Am. Assoc. Adv. Sci. **286**(5439), 509–512 (1999)
7. Batagelj, V., Zaversnik, M.: An O(m) algorithm for cores decomposition of networks (2003). arXiv:cs/0310049
8. Bhawalkar, K., Kleinberg, J., Lewi, K., Roughgarden, T., Sharma, A.: Preventing unraveling in social networks: the anchored k-core problem. SIAM J. Discrete Math. **29**(3), 1452–1475 (2015)
9. Capocci, A., et al.: Preferential attachment in the growth of social networks: the internet encyclopedia wikipedia. Phys. Rev. E **74**(3), 036116 (2006)
10. Chhabra, S.S., Brundavanam, A., Shannigrahi, S.: An alternative explanation for the rise and fall of MySpace (2014). arXiv:1403.5617
11. Dorogovtsev, S.N., Mendes, J.F.F.: Scaling behaviour of developing and decaying networks. EPL (Europhys. Lett.) **52**(1), 33 (2000)
12. Garcia, D., Mavrodiev, P., Schweitzer, F.: Social resilience in online communities: the autopsy of Friendster. In: Proceedings of the First ACM Conference on Online Social Networks, pp. 39–50. ACM (2013)
13. Iwata, S., Fleischer, L., Fujishige, S.: A combinatorial strongly polynomial algorithm for minimizing submodular functions. J. ACM (JACM) **48**(4), 761–777 (2001)
14. Jackson, M.O.: A survey of network formation models: stability and efficiency. In: Group Formation in Economics: Networks, Clubs, and Coalitions, pp. 11–49 (2003)
15. Jin, E.M., Girvan, M., Newman, M.E.: Structure of growing social networks. Phys. Rev. E **64**(4) (2001)
16. Kairam, S.R., Wang, D.J., Leskovec, J.: The life and death of online groups: predicting group growth and longevity. In: Proceedings of the Fifth International Conference on Web Search and Data Mining, pp. 673–682. ACM (2012)
17. Kordestani, A.A., Limayem, M., Salehi-Sangari, E., Blomgren, H., Afsharipour, A.: Why a few social networking sites succeed while many fail. In: The Sustainable Global Marketplace, pp. 283–285. Springer (2015)
18. Kossinets, G., Watts, D.J.: Empirical analysis of an evolving social network. Science **311**(5757), 88–90 (2006)
19. Krause, A., Golovin, D.: Submodular function maximization. In: Tractability: Practical Approaches to Hard Problems (2012)
20. Kumar, R., Novak, J., Tomkins, A.: Structure and evolution of online social networks. In: Proceedings of the 12th ACM SIGKDD International Conference on Knowledge Discovery and Data Mining, pp. 611–617 (2006)
21. Leskovec, J., Kleinberg, J., Faloutsos, C.: Graphs over time: densification laws, shrinking diameters and possible explanations. In: Proceedings of the Eleventh ACM SIGKDD, pp. 177–187. ACM (2005)
22. Malliaros, F.D., Vazirgiannis, M.: To stay or not to stay: modeling engagement dynamics in social graphs. In: Proceedings of the 22nd ACM International Conference on Conference on Information and Knowledge Management, CIKM'13, pp. 469–478. ACM, New York, NY, USA (2013)
23. Mislove, A., Koppula, H.S., Gummadi, K.P., Druschel, P., Bhattacharjee, B.: Growth of the flickr social network. In: Proceedings of the First Workshop on Online Social Networks, pp. 25–30. ACM (2008)
24. Nemhauser, G.L., Wolsey, L.A.: Best algorithms for approximating the maximum of a submodular set function. Math. Oper. Res. **3**(3), 177–188 (1978)

25. Newman, M.E.: Clustering and preferential attachment in growing networks. Phys. Rev. E **64**(2), 025102 (2001)
26. Ribeiro, B.: Modeling and predicting the growth and death of membership-based websites. In: Proceedings of the 23rd International Conference on World Wide Web, pp. 653–664. ACM (2014)
27. Stieger, S., Burger, C., Bohn, M., Voracek, M.: Who commits virtual identity suicide? Differences in privacy concerns, internet addiction, and personality between facebook users and quitters. Cyberpsychol. Behav. Soc. Netw. **16**(9), 629–634 (2013)
28. Torkjazi, M., Rejaie, R., Willinger, W.: Hot today, gone tomorrow: on the migration of MySpace users. In: Proceedings of the 2nd ACM Workshop on Online Social Networks, pp. 43–48. ACM (2009)

Part IV
Applications of Network Science

Complex Reaction Network in Silane Plasma Chemistry

Yasutaka Mizui, Kyosuke Nobuto, Shigeyuki Miyagi and Osamu Sakai

Abstract Chemical reactions become significantly complex when plasma is introduced in a reaction space. We study silane plasma chemistry, and centrality indices derived from the reaction network indicate several points of information about species in reactions as well as macroscopic topology in the entire network graph. Stable species, unstable species and electrons play different roles as triggers or products of reactions, and this analytical method provides several points that cannot be revealed by rate-equation calculations, which have been popular in chemical analysis.

1 Introduction

Chemical reactions have been studied using network analysis for several decades since their reactions are usually successive and complicated [1, 2]. It is so essential to elucidate roles of species in reactions like products, subproducts and intermediates that many scientific and technological approaches to their estimations have been performed, usually by numerical calculations using rate equations [3, 4]. In particular, chemical reactions in low-temperature plasmas are more complicated than those in other reactions schemes [3, 4]. Then, such chemical reactions may form a complex network which include various statistical and/or topological properties [5].

Not to estimate precise densities of chemical species but to obtain network properties in a macroscopic point of view, we recently demonstrated network analysis of the directed graphs for methane plasma chemistry [6]. We verified the particular roles of electrons, which are origins of chemical activities in plasma chemistry, and clarified a role of each species using a centrality index based on eigenvector centrality measures. Further studies on other species, which may include larger number of reactions, will open possibilities to confirm validity of analyses and create suitable measures as well as collect various data for accomplishing database for future dictionary learning.

Y. Mizui · K. Nobuto · S. Miyagi · O. Sakai (✉)
The University of Shiga Prefecture, 2500 Hassaka-cho, Hikone, Shiga 522-8533, Japan
e-mail: sakai.o@e.usp.ac.jp

B. Gonçalves et al. (eds.), *Complex Networks VIII*,
Springer Proceedings in Complexity, DOI 10.1007/978-3-319-54241-6_11

Here we consider silane (SiH_4) about its decomposition and subsequent reactions such as ionization and polymerization. In Ref. [3], very complicated reactions are analyzed numerically by solving more than 100 rate equations implicitly with reaction rate constants. The results are quite precise to predict densities of product species, which is a kind of visualization about outlooks of each species, although it is not easy at a glance to recognize role(s) and/or function(s) of each species in a reaction network, such as final products, subproducts, or intermediates.

In this report, we analyze chemical reactions of silane as a mother gas in plasma chemistry. The numbers of reactions and species are larger than those of methane [6], and we recognize several common and different points. Chemically stable species play different roles due to their mother atoms. Electrons and unstable species are also analyzed using eigenvector centrality measures.

2 Analytical Methods and Results

2.1 Analytical Methods

Figure 1 displays the entire network of chemical reactions in silane plasma described in Ref. [3]. The number of nodes is 58, and that of edges which represent reactions is 222. Nodes and edges are defined as shown in Table 1. In plasma chemistry investigated here, all reactions are assumed to be irreversible.

Among the 222 reactions, we show typical examples used in this study in the following.

$$e^- + SiH_4 \longrightarrow SiH_3^+ + H + 2e^-. \tag{1}$$

$$SiH_4 + H \longrightarrow SiH_3 + H_2. \tag{2}$$

$$SiH_3^+ + SiH_4 \longrightarrow Si_2H_5^+ + H_2. \tag{3}$$

$$e^- + SiH_3^+ \longrightarrow SiH_2 + H. \tag{4}$$

Reaction (1) is dissociative ionization of the mother gas SiH_4 by electron impact. Reaction (2) was already shown in Table 1, and H atoms generated in Reaction (1) induce dissociation of SiH_4. Generated ions frequently yield larger ions in reactions like Reaction (3), and electrons and ions recombine according to Reaction (4), for instance.

We calculate values of *simplified pagerank index* as a centrality index. These are elements of eigenvector centrality measures using the modified adjacency matrix whose zero elements are changed to small non-zero values; this centrality index is simplified from the original one proposed as PageRank [7].

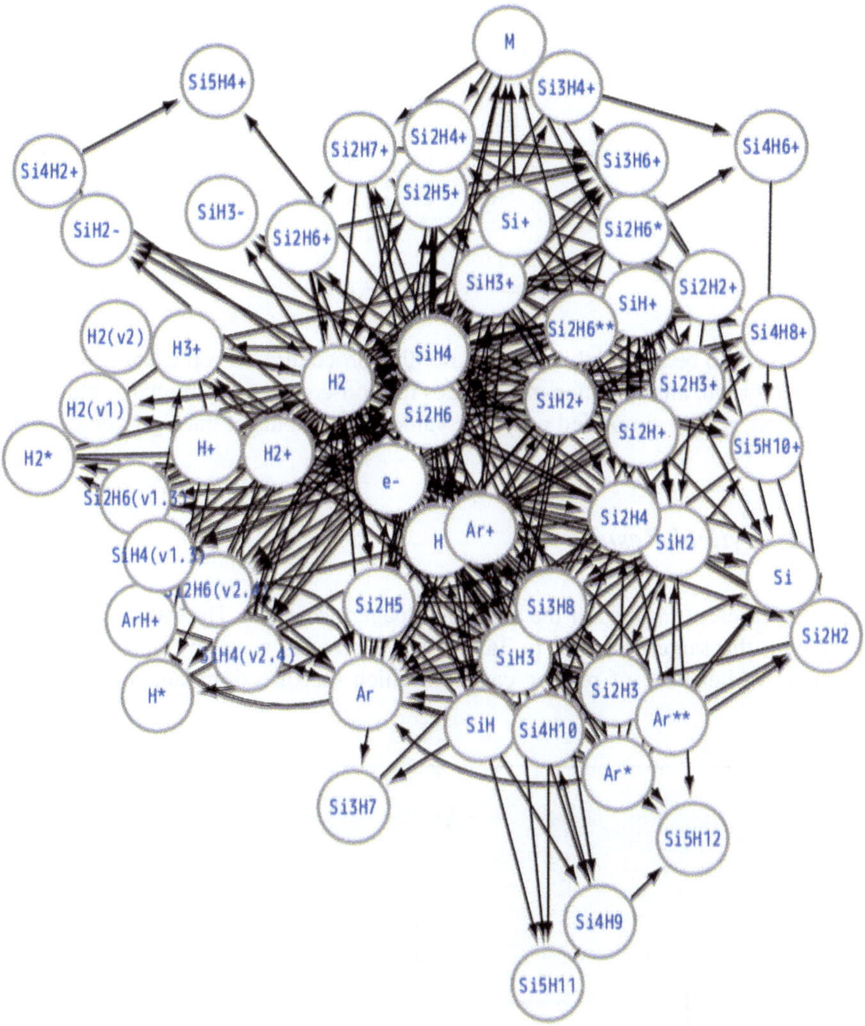

Fig. 1 Graph of reaction network analyzed in silane plasma chemistry. Reactions are listed in Ref. [3]

Table 1 Example of reaction and resultant nodes and edges

Reaction	$SiH_4 + H \rightarrow SiH_3 + H_2$
Nodes	Node a: SiH_4, Node b: H, Node c: SiH_3, Node d: H_2
Edges	Edge A: from a to c
	Edge B: from a to d
	Edge C: from b to c
	Edge D: from b to d

The calculation procedure is as follows. The adjacency matrix **A** is converted into the target matrix **M** via transition probability matrix $\mathbf{A}'$; all of them are $n \times n$ matrices. When we denote the entry of **A** on the i-th row and on the j-th column as A_{ij},

$$A'_{ij} = \frac{A_{ij}}{\sum_j A_{ij}}, \tag{5}$$

$$M_{ij} = cA'_{ji} + (1 - c)\frac{1}{n}, \tag{6}$$

where c is set to be 0.85 in this study. After this procedure, **M** represents a strongly-connected graph, although non-zero values for non-connectivity branch remain very small. Then, we get simplified pagerank values as elements of the eigenvector of **M**.

2.2 Analytical Results

Figure 2 represents values of the simplified pagerank index. The values of forward cases with the same direction in Table 1 show effects caused by other species as products, while those of reverse cases in which the directions are reversed from

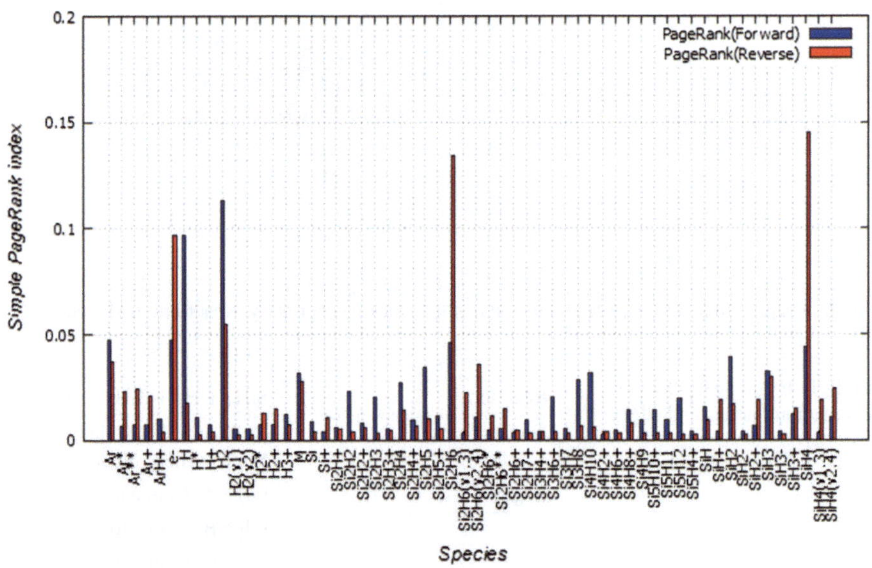

Fig. 2 Simplified pagerank index for species as nodes in network. Symbols used for species are according to notation in Ref. [3]; '*' indicates excited states, and 'v' denotes vibrationally-excited states. 'M' is an arbitrary species which work as a collision partner for specific reactions

those in Table 1 indicate influences on other species as triggers of reactions. Among stable species, Si-related ones (SiH_4, Si_2H_6, and so forth) are influential on other species, while H-related one (H_2) is less influential and rather affected by other species. Electrons are significantly influential as well, similar to the case of methane. Except stable species, SiH_3 is the most important species as an influential species and simultaneously affected one. That is, SiH_3 is a key species as a substantial intermediate in this network web, similar to CH_3 in methane plasma chemistry [6].

As described in Ref. [1], chemical reactions have been comprehended using graphs for decades, but they are simple and usually in chain-like structures [2]. So far, since number of nodes in such graphs are limited in usual liquid and gas chemistry (without electrons), there have been no requirements on analysis based on complex networks [5]. However, as we see in Fig. 1, the reaction network in plasma chemistry is so complex, and this kind of analysis will contribute to understand unclear roles of each species, leading to further efficient chemical processes via future optimization using this kind of method.

3 Conclusion

We analyzed chemical reactions of silane as a mother gas in plasma chemistry. The numbers of reactions and species are larger than those of methane [6], and we recognized several common and different points of complex networks in plasma chemistry. This study reveals that chemically stable species play different roles due to their mother atoms. Electrons and unstable species are also analyzed using an eigenvector centrality measure, and we found that SiH_3 in silane plasmas plays the similar role to CH_3 in methane plasmas.

Acknowledgements One of the authors (OS) thanks Prof. M. J. Kushner at University of Michigan, Prof. T. Murakami at Seikei University, Prof. S. Nunomura at National Institute of Advanced Industrial Science and Technology, and Prof. T. Akiyama at The University of Shiga Prefecture for their useful comments on this study. This study was supported by Grant-in-Aid for Scientific Research from the Japanese Ministry of Education, Culture, Sports, Science and Technology, Japan.

References

1. Temkin, O.N., Zeigarnik, A.V., Bonchev, D.: ChemicaL Reaction Networks. CRC Press, Boca Raton (1996)
2. Gorban, A.N., Yablonsky, G.S.: Extended detailed balance for systems with irreversible reactions. Chem. Eng. Sci. **66**, 5388–5399 (2011)
3. Kushner, M.J.: A model for the discharge kinetics and plasma chemistry during plasma enhanced chemical vapor deposition of amorphous silicon. J. Appl. Phys. **63**, 2532–2551 (1988)
4. Murakami, T., Niemi, K., Gans, T., O'Connell, D., Graham, W.G.: Chemical kinetics and reactive species in atmospheric pressure helium-oxygen plasmas with humid-air impurities. Plasma Sources Sci. Technol. **22**, 015003-1–22 (2013)

5. Albert, R., Barabasi, A.-L.: Statistical mechanics of complex networks. Rev. Mod. Phys. **74**, 47–97 (2002)
6. Sakai, O., Nobuto, K., Miyagi, S., Tachibana, K.: Analysis of weblike network structures of directed graphs for chemical reactions in methane plasmas. AIP Adv. **5**, 107140-1–6 (2015)
7. Brin, S., Page, L.: The anatomy of a large-scale hypertextual Web search engine. Comp. Netw. ISDN Syst. **30**, 107–117 (1998)

Seeing Red: Locating People of Interest in Networks

Pivithuru Wijegunawardana, Vatsal Ojha, Ralucca Gera and Sucheta Soundarajan

Abstract The focus of the current research is to identify people of interest in social networks. We are especially interested in studying dark networks, which represent illegal or covert activity. In such networks, people are unlikely to disclose accurate information when queried. We present REDLEARN, an algorithm for sampling dark networks with the goal of identifying as many nodes of interest as possible. We consider two realistic lying scenarios, which describe how individuals in a dark network may attempt to conceal their connections. We test and present our results on several real-world multilayered networks, and show that REDLEARN achieves up to a 340% improvement over the next best strategy.

Keywords Multilayered networks · Sampling · Lying scenarios · Nodes of interest

1 Introduction and Motivation

Today's complex environment requires decision makers to act in an overwhelmingly rich network environment, often based on partial information of that network. It is often desirable to locate "people of interest" (POI) residing in such networks while

Ralucca Gera would like to thank the DoD for partially sponsoring the current research.

P. Wijegunawardana (✉) · S. Soundarajan
Department of Electrical Engineering & Computer Science,
Syracuse University, New York, USA
e-mail: ppwijegu@syr.edu

S. Soundarajan
e-mail: susounda@syr.edu

V. Ojha
Dougherty Valley High School, San Ramon, CA, USA
e-mail: vatsalojha@gmail.com

R. Gera
Department of Applied Mathematics, Naval Postgraduate School, Monterey, CA, USA
e-mail: RGera@nps.edu

B. Gonçalves et al. (eds.), *Complex Networks VIII*,
Springer Proceedings in Complexity, DOI 10.1007/978-3-319-54241-6_12

they conceal themselves or others. Our work was motivated by study of terrorist networks, which can be modeled multilayered networks where each layer is defined by a different relationship (e.g., relationships indicate organizations these terrorists belong to, the schools or trainings they went to, kinship, recruiting and so on.)

In this paper, we consider the goal of sampling a 'dark' network (i.e., a network representing illegal or covert activity) in such a way that we observe as many POIs as possible. We present REDLEARN, a novel learning-based algorithm for sampling networks with the goal of finding as many POIs as possible. We show that in cases where the POIs exhibit homophily (i.e., are likely to be connected to other POIs), a simple strategy of choosing the node with the most POI neighbors works well. However, in the more realistic scenario where POIs hide their connections with other POIs, REDLEARN shows outstanding performance, beating the next best strategy by up to 340%.

Problem Definition: We refer to nodes representing POIs as 'red' nodes, and other nodes as 'blue', giving us a purple network. We assume that there is an unobserved, underlying graph $G = (V, E)$, in which each node $v \in V$ has color $C_v \in \{red, blue\}$. We begin with having knowledge of only one red node in G.

To increase our observation of the network, we place *monitors* on nodes. A monitor tells us (1) the true color of the node being placed on, (2) the true neighbors of that node, and (3) the colors of the node's neighbors, possibly with inaccuracies. For example, placing a monitor on a suspected terrorist could represent determining whether that person is actually a terrorist, determining who his or her e-mail or phone contacts are, and questioning the individual about whether those neighbors are themselves terrorists. Naturally, some individuals may lie about the colors of those neighbors.[1]

We assume that we are given a budget of b monitors, and can place those monitors on any node that has been observed. In the first step, we must place a monitor on the initially observed node. We then place a monitor on any node that has been observed as a neighbor of a previously-monitored node.

Related Work: Our work is related to work on analyzing dark networks, a special type of social network [4]. A dark network is network that is illegal and covert [14], whose members are actively trying to conceal network information even at the expense of efficiency [4], and the existing connections are used infrequently [14]. Because a dark network is deceptive by nature, we examine the lying methodologies along with the discovery methods in looking for the POI.

There are a multitude of sampling techniques for network exploration, including random walks [3, 11, 13], biased random walks [9], or walks combined with reversible Markov Chains [2], Bayesian methods [8], or standard exhaustive search algorithms like depth-first or breadth-first searches, such as [1, 5–7, 12]. However, these methods generally do not use node attributes.

[1] We consider two realistic 'lying scenarios'; these are described in Sect. 2.1.

2 Proposed Method: REDLEARN

A monitor placement strategy is an incremental sampling strategy. A *monitored node* is a node with a monitor placed on it. At each step, the placement of the next monitor is determined based on the observed topology of the graph, known colors of nodes (observed by monitors placed directly on those nodes), and the stated colors of monitored nodes' neighbors (i.e., for each neighbor of a monitored node, whether the monitored node said that neighbor was red or blue).

We now describe several natural monitor placement strategies as comparison algorithms in our experiments.

Smart Random Sampling (SR): In each step, the Smart Random Placement strategy places a monitor on a random unmonitored node.

Red Score (RS): If a node v reports its neighbor u as red, the score associated with node u is increased by one, making it more suspicious. This strategy selects the node with highest red score to place the next monitor.

Most Red Say Red (MRSR): The MRSR strategy places a monitor on the node with the greatest number of red neighbors who report it as a red node.

Most Red Neighbors (MRN): The MRN placement strategy places a monitor on the node with the most known red neighbors. This strategy would likely work best in a network with high homophily.

2.1 REDLEARN*: A Learning Based Monitor Placement Strategy*

When determining which node v to place the next monitor on the strategies above consider the colors of v's neighbors and/or the color that each of v's monitored neighbors reported, the presence of homophily, and the reported color of the neighbors.

To overcome these dependencies, we propose REDLEARN, a learning based monitor placement strategy. Our goal is to predict the probability of a node v being red ($P(v = R)$) based on the observed network structure and what v's neighbors say about v. We model this as a two class classification problem, but rather than looking at the assigned label (Red or Blue), we are more interested in finding $P(v = R)$. Once these probabilities are determined, REDLEARN places the next monitor on the node with the highest such probability.

Features: Table 1 describes the set of features used in our learning based monitor placement algorithm. There are two types of features: (*a*) Network structure-based features (1, 2, 3), and (*b*) Neighbor answer-based features (4, 5, 6, 7, 8).

Inferred Probability of Being Red: We formulate four different probabilities to measure the trustworthiness of colors given by differently colored nodes (i.e., whether a monitored node lies or is honest about its neighbors' colors). Consider a node v which was discovered through a monitor placed on node u. Equation 1 shows how to calculate $P(v = R|color(u) \wedge color(u\ says\ v))$ when $v = R$, $u = R$ and u says v is

Table 1 Classification features for REDLEARN. Consider a node v with neighbors $N(v)$

	Feature	Description
(1)	Number of Red Neighbors	$\lvert\{u \in N(v) \vert c_u = R\}\rvert$
(2)	Number of Blue neighbors	$\lvert\{u \in N(v) \vert c_u = B\}\rvert$
(3)	Number of Red triangles if v is red	$\{u, w \in N(v) \vert u \in N(w) \cap w \in N(u) \cap c_u = c_w = R\}$
(4)	Red score	$\lvert\{u \in N(v) \vert (u\ says\ R)\}\rvert$
(5)	Number of Red neighbors saying red	$\lvert\{u \in N(v) \vert (u\ says\ R) \cap c_u = R\}\rvert$
(6)	Number of red neighbors saying blue	$\lvert\{u \in N(v) \vert (u\ says\ B) \cap c_u = R\}\rvert$
(7)	Number of blue neighbors saying red	$\lvert\{u \in N(v) \vert (u\ says\ R) \cap c_u = B\}\rvert$
(8)	Number of blue neighbors saying blue	$\lvert\{u \in N(v) \vert (u\ says\ B) \cap c_u = B\}\rvert$
(9)	Inferred probability of being red	$P^I(v = R)$

red. Other probabilities can be calculated by changing components of this equation as appropriate.

$$P(v = R \vert (u = R) \wedge (u\ Says\ R)) = \frac{\lvert\{(v = R) \cap (u = R) \cap (u\ says\ R)\}\rvert}{\lvert\{(u = R) \cap (u\ says\ R)\}\rvert} \tag{1}$$

Given a node v, we calculate the inferred probability, $P^I(v = R)$ using Eq. 2.

$$P^I(v = R) = \frac{\sum_{u \in N(v)} P(v = R \vert color(u) \wedge color(u\ says\ v))}{\lvert N(v)\rvert} \tag{2}$$

The training data for this classification problem comes from the monitors placed so far and observed true colors. We predict $P(v = R)$ for each unmonitored node. We use logistic regression as the classification algorithm in our experiments.

Algorithm 1 Learning based monitor placement

```
procedure LEARNING(start,budget)
    G ← Graph
    G.add(start), G.add(N(start))                    ▷ Starting node and neighbors
    while budget>0 do
        Monitors ← list of monitored nodes in G
        TrainingData ← feature vectors for Monitors
        Train classifier using TrainingData
        NotMonitors ← list of not yet monitored nodes in G
        for v ∈ NotMonitors do
            Get feature vector for v
            P(v=R) ← predict v's probability of Red using learning model
        end for
        Choose node v with maximum P(v = R) from NotMonitors
        budget ← (budget − 1)
        Use v as next monitor
    end while
end procedure
```

3 Experimental Set up

3.1 Datasets

PokeC Network: The PokeC network is part of a Slovenian online social network.[2] Each node has some number of associated user attributes (e.g., age, region, gender, interests, height etc.). We use a sample of this network containing all nodes in the region "kosicky kraj, michalovce". This sampled network contains 26, 220 nodes and 241, 600 edges. We assign node colors based on two different node attributes: *age* (a node with age in the range 28–32 is marked red, and blue otherwise, giving 1736 red nodes) and *height* (a user of height less than 160 *cm* is marked red, giving 1668 red nodes).

Noordin Top Network is a terrorist network with 139 nodes and 1042 edges depicting several types of relationships between them ('Noordin Top' is the name of the leader of this network) [10].[3] In this network, every node is a terrorist, and POIs are those who communicate using some particular communication medium. We have identified five different communication mediums, and label nodes that use them as POIs: electronic (9 red nodes), print media (5 red nodes), support materials (9 red nodes), video (11 red nodes) and communication medium unknown (18 red nodes).

Both networks have high homophily for red nodes (red nodes tend to be connected to each other). However, in a dark network where red nodes are actively trying to hide their presence, these nodes would conceal the existence of such connections (for example, instead of using their normal cell phone to make calls to other red nodes, a red node might use a burner phone for such calls). To account for this, we also consider versions of our datasets where all connections between red nodes are removed. Note that this type of network presents a much more challenging setting, as one cannot simply rely on homophily to find red nodes.

3.2 Lying Scenarios

In absence of ground truth, we formulate lying scenarios: we assume the existence of a hierarchy among the nodes, where nodes are more likely to lie to protect those above them in the hierarchy. We assume that the red nodes are fully aware of the hierarchy, blue nodes may or may not be aware, and that nodes may lie not only about the color of red nodes (i.e., lie to protect POIs), but also about the color of blue nodes (i.e., as a distraction).

Consider nodes u and v, where $u, v \in Edges$. The probability that u lies about v, $P(u \text{ } lie \text{ } v)$ depends on: (1) The color of u (C_u) and color of v (C_v),(2) The inherent

[2]Obtained from http://snap.stanford.edu/data/.

[3]Obtained from https://sites.google.com/site/sfeverton18/research/appendix-1.

Table 2 The probability that node u lies about node v's color $P(u\ lie\ v)$ depending on u's and v's colors and lying scenarios

	LS1:Blue nodes know about red nodes		LS2:Blue nodes dont know about red nodes	
U/V	**Red**	**Blue**	**Red**	**Blue**
Red	Eq. 3	Eq. 4	Eq. 3	Eq. 4
Blue	Eq. 3	Eq. 4	1.0	0.0

honesty of u (H_u), where higher H values indicate that u is more predisposed to telling the truth and (3) The hierarchical position of u (L_u) relative to the position of v (L_v).
The probability u will lie about a red node: where $\frac{L_v}{L_u}$ indicates how far above v is in the hierarchy compared to u and $1 - H_u$ is probability that u will lie.

$$P(u\ lie\ v|v = Red) = \min\{(1 - H_u) * \frac{L_v}{L_u}, 1\} \tag{3}$$

The probability u will lie about a blue node depends on u's honesty and is calculated as $(1 - H_u)$:

$$P(u\ lie\ v|v = Blue) = (1 - H_u) \tag{4}$$

We perform 25 runs of each monitor placement strategy, varying the honesty assignment and the colors that nodes say about neighbors between runs. In each run, we begin with a randomly selected red node and we consider budgets up to half the number of nodes in the network.

The honesty of each node is drawn from a normal distribution, $h \sim \mathcal{N}(0.5, 0.125)$. In the Noordin Top network, the ground truth hierarchy scores are Strategist (score 5),Commander; Religious Leader (score 4),Trainer/instructor; Bomb maker; Facilitator; Propagandist; Recruiter (score 3), Bomber/fighter; Suicide Bomber; Courier; Recon/Surveillance (score 2) and unknown (score 1). In the PokeC network, we set the hierarchy score to be the degree of the node.

Given a particular lying scenario, a monitored node u lies about a neighbor v's color with probability $P(u\ lie\ v)$ as shown in Table 2.

4 Results and Analysis

As an example, Fig. 1 shows results on the NoordinComs4 network with edges between red nodes (left two plots) and without (right two plots). When there is homophily, the problem becomes easy, and the simple strategy of monitoring the node with the most red neighbors (MRN) is best. However, note that in both lying scenarios, RedLearn is close behind the MRN strategy (because it needs time to train, it doesn't quite match the performance of MRN). However, we see from the

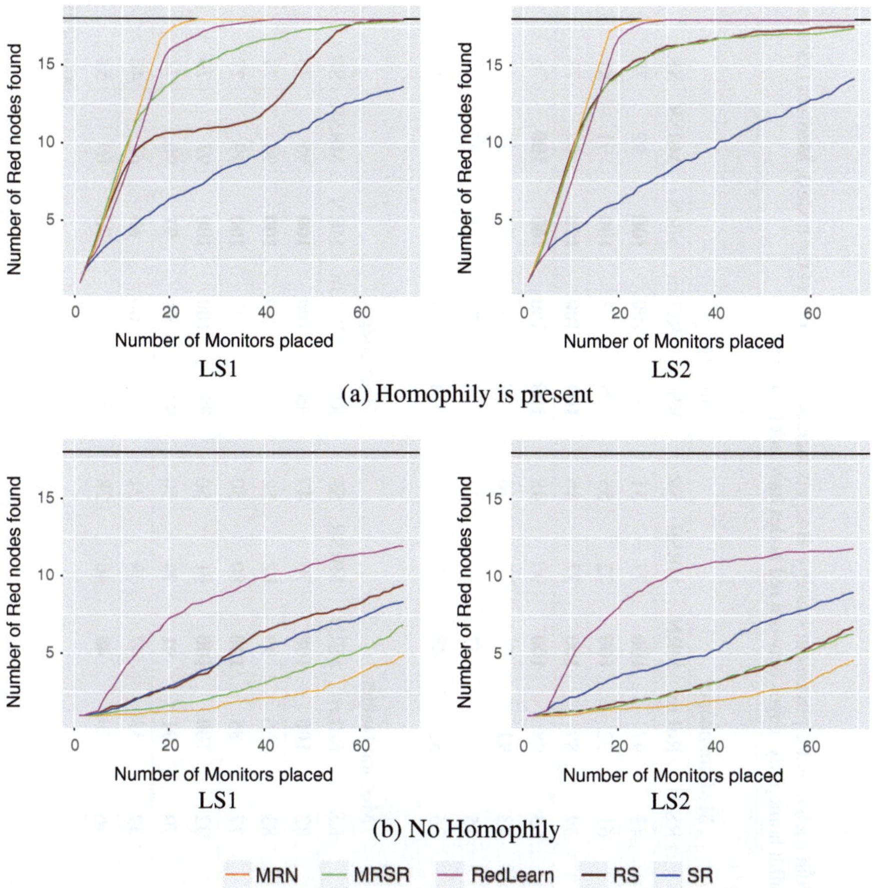

Fig. 1 Comparison of monitor placement strategies on the NoordinComs4 network. LS1:All nodes aware of *red* nodes. LS2: Only *red* nodes aware of *red* nodes. The *black* line indicates the total number of *red* nodes present in the network

right two figures that when edges between red nodes are removed, the MRN strategy performs very poorly. In this setting, REDLEARN performs much better than all comparison methods: it is able to learn the patterns and structural characteristics of red nodes, and by incorporating what neighbors say about a node, achieves strong performance.

Due to space constraints, we summarize results by showing the percentage of red nodes found from each monitor placement strategy for other networks in Table 3. We see similar patterns across all networks: when there are edges between red nodes, it is enough to select the node with the most red neighbors; but when these edges are concealed, REDLEARN is the clear winner. Even when there are edges between red nodes, REDLEARN usually achieves performance close to the MRN strategy.

Table 3 Comparison of the percentage of red nodes found from each monitor placement strategy. Budgets include Low (10% of the nodes), Medium (25% of the nodes), and High (50% of the nodes). These networks exhibit homophily: edges between red nodes have not been removed

Lying Scenario 1, original data (with homophily)

	Low budget					Medium budget					High budget				
Network/ Strategy	RS	RdLrn	MRN	MRSR	SR	RS	RdLrn	MRN	MRSR	SR	RS	RdLrn	MRN	MRSR	SR
NrdnComs1	28	74	**97**	52	32	43	97	**100**	77	51	97	**100**	**100**	92	75
NrdnComs2	42	37	**62**	48	42	61	72	**100**	72	55	99	93	**100**	91	81
NrdnComs3	33	63	**83**	52	27	59	89	**100**	77	46	**100**	**100**	**100**	97	73
NrdnComs4	54	60	**70**	66	28	63	98	**100**	90	48	**100**	**100**	**100**	**100**	76
NrdnComs5	34	67	**84**	52	32	43	**91**	**91**	75	46	88	**91**	**91**	86	67
PokeC age	5	14	**22**	20	7	15	43	**47**	39	21	48	**73**	68	62	47
PokeC height	14	14	21	**23**	11	36	32	**48**	47	28	**74**	64	73	69	54

Lying Scenario 2, original data (with homophily)

	Low budget					Medium budget					High budget				
Network/ Strategy	RS	RdLrn	MRN	MRSR	SR	RS	RdLrn	MRN	MRSR	SR	RS	RdLrn	MRN	MRSR	SR
NrdnComs1	57	78	**89**	57	33	82	**100**	**100**	79	47	96	**100**	**100**	95	73
NrdnComs2	56	54	**83**	55	38	83	66	**99**	82	52	94	89	**100**	94	74
NrdnComs3	50	70	**76**	52	34	77	85	**100**	80	50	99	97	**100**	99	77
NrdnComs4	68	62	**74**	67	28	92	**100**	**100**	91	50	98	**100**	**100**	97	79
NrdnComs5	59	64	**88**	59	32	79	**91**	**91**	79	50	89	**91**	**91**	90	74
PokeC age	20	12	**22**	19	7	39	33	**47**	39	21	62	60	**68**	62	46
PokeC height	**23**	12	21	**23**	12	46	29	**48**	46	28	69	62	**73**	69	54

(continued)

Table 3 (continued)

Lying Scenario 1, no homophily															
	Low Budget					Medium Budget					High Budget				
Network/ Strategy	RS	RdLrn	MRN	MRSR	SR	RS	RdLrn	MRN	MRSR	SR	RS	RdLrn	MRN	MRSR	SR
NrdnComs1	16	**33**	12	14	22	38	**46**	20	27	36	**72**	64	40	48	58
NrdnComs2	30	**70**	21	26	35	50	**82**	26	41	55	86	**94**	52	63	84
NrdnComs3	21	**59**	12	14	22	57	**82**	16	28	44	76	**99**	41	57	76
NrdnComs4	12	**30**	6	8	12	31	**52**	11	15	28	53	**67**	28	38	47
NrdnComs5	13	**35**	10	11	16	30	**51**	12	18	26	52	**55**	28	34	40
PokeC age	5	**13**	5	6	7	14	**34**	16	18	20	43	**59**	39	41	44
PokeC height	13	**14**	5	7	11	**33**	**33**	15	19	27	**69**	59	37	44	52

Lying Scenario 2, no homophily															
	Low Budget					Medium Budget					High Budget				
Network/ Strategy	RS	RdLrn	MRN	MRSR	SR	RS	RdLrn	MRN	MRSR	SR	RS	RdLrn	MRN	MRSR	SR
NrdnComs1	14	**33**	12	14	22	21	**53**	19	24	41	53	**64**	40	48	63
NrdnComs2	26	**58**	22	25	37	40	**70**	26	35	54	68	83	54	70	**84**
NrdnComs3	15	**64**	12	16	23	26	**85**	17	23	38	54	**98**	41	57	70
NrdnComs4	8	**35**	7	8	15	15	**59**	10	15	27	38	**66**	26	35	50
NrdnComs5	9	**39**	9	11	18	16	**47**	13	17	27	33	**53**	27	35	42
PokeC age	6	**10**	5	6	7	16	**26**	14	16	18	42	**59**	39	41	44
PokeC height	6	**12**	5	7	10	19	**28**	15	19	26	43	**58**	37	43	52

5 Conclusions and Further Directions

By nature, members of dark networks conceal information, but while deceptive and sparse, these networks are still structured. To exploit these properties, we created RedLearn, a learning-based method for locating People of Interest in dark networks. RedLearn uses features from simpler methods and learns how to identify red nodes in networks. We showed that RedLearn outperforms the other methods in cases where one cannot rely on homophily to identify red nodes.

In our future work, one interesting direction is to consider the dynamicity of the network (both on the edge and node rate of birth and retirement), as well as a more sophisticated model of the concealed nodes and relationships.

References

1. Adamic, L.A., Lukose, R.M., Puniyani, A.R., Huberman, B.A.: Search in power-law networks. Phys. Rev. E **64**(4), 046135 (2001)
2. Aldous, D., Fill, J.: Reversible markov chains and random walks on graphs (2002)
3. Asztalos, A., Toroczkai, Z.: Network discovery by generalized random walks. EPL (Europhysics Letters) **92**(5), 50008 (2010)
4. Baker, W.E., Faulkner, R.R.: The social organization of conspiracy: illegal networks in the heavy electrical equipment industry. Am. Sociol. Rev. pp. 837–860 (1993)
5. Biernacki, P., Waldorf, D.: Snowball sampling: problems and techniques of chain referral sampling. Soc. Methods Res. **10**(2), 141–163 (1981)
6. Bliss, C.A., Danforth, C.M., Dodds, P.S.: Estimation of global network statistics from incomplete data. PloS one **9**(10), e108471 (2014)
7. Davis, B., Gera, R., Lazzaro, G., Lim, B.Y., Rye, E.C.: The marginal benefit of monitor placement on networks. In: Complex Networks VII, pp. 93–104. Springer (2016)
8. Friedman, N., Koller, D.: Being bayesian about network structure. a bayesian approach to structure discovery in bayesian networks. Mach. Learn. **50**(1–2), 95–125 (2003)
9. Fronczak, A., Fronczak, P.: Biased random walks in complex networks: the role of local navigation rules. Phys. Rev. E **80**(1), 016107 (2009)
10. Gera, R., Miller, R., MirandaLopez, M., Warnke, S.: Developing multilayered dark networks to enhance community identification. Submitted for publication (2016)
11. Hughes, B.D.: Random walks and random environments (1996)
12. Leskovec, J., Faloutsos, C.: Sampling from large graphs. In: SIGKDD, pp. 631–636. ACM (2006)
13. Noh, J.D., H. Rieger: Random walks on complex networks. Phys. Rev. Lett. **92**(11), 118701 (2004)
14. Raab, J., Milward, H.B.: Dark networks as problems. J. Public Adm. Res. Theory **13**(4), 413–439 (2003)

Understanding Subject-Based Emoji Usage Using Network Science

S.M. Mahdi Seyednezhad and Ronaldo Menezes

Abstract The use of "Emoticons" and "Emojis" in social media as well as most online writing has become the *de-facto* standard on how to express emotions, feelings, etc. Although there are more that 1,000 emojis, not much has been done to understand the way in which people use these characters. The large set of emojis available brings two questions: (i) How can users make full use of the emojis available? and (ii) Would it be possible to build a recommendation system for emoji usage in text? This paper moves towards a greater understanding of emoji usage by mapping possible relations between these special characters in common text. We look at possible regularities in emoji usages in written, subject-specific, text corpora. We build co-occurrence networks of emoji based on two datasets and show that the structure of these networks are not random and more like a truncated power-law, but more interesting, we show that the structure has similar characteristics despite the text being subject-specific.

Keywords Emoji · Word co-occurrence networks · Network science · Twitter data

1 Introduction

Our inability to express emotions in written language is notorious. For instance, who has never tried to send an email with some sarcasm and found that it was not well understood. The misunderstanding arises because the text does not convey your facial expressions or perhaps your tone of voice; crucial for sarcasm. In 1982, Scott Fahlman, a professor at Carnegie Mellon University (CMU) proposed what is considered the first use of a emoticon in a message to a general CMU mailing list.

After the first use, the idea of emoticons spread quickly and many variations have been proposed by Fahlman and others and today emoticons are still commonly used.

S.M.M. Seyednezhad (✉) · R. Menezes
BioComplex Laboratory, School of Computing, Florida Institute of Technology, Melbourne, USA
e-mail: sseyednezhad2013@my.fit.edu

R. Menezes
e-mail: rmenezes@fit.edu

B. Gonçalves et al. (eds.), *Complex Networks VIII*,
Springer Proceedings in Complexity, DOI 10.1007/978-3-319-54241-6_13

Emojis were introduced in 2010 in Unicode 6.0 and today there are 1,088 emojis defined in Unicode 9.0. These emojis are graphical version of the emoticons and include representations such as 👕, 🍺, 🏃, and 😃. With the growth in the number of Internet users, the need for the emojis has been risen. The variety of emojis correspond to the diversity of emotional feelings in humans [8] but it also grew to other usages such as flags, animals, symbols, activities, etc.

Despite their popularity (e.g. emoji are used in nearly 800% more campaigns than in 2015[1]), there has been little movement on trying to understand how society uses these emojis. Even though,"emojis won the battle of words" as claimed by the New York Times,[2] their use relies completely on user knowledge about a particular instance of the characters. The popularity of emojis has lead the Oxford dictionary to select the word of 2015 as "face with tears of joy" which is the name of an emoji (😂).

Another interesting aspect about the emoji phenomena is that they become akin to a universal language because many are understood similarly in different locations easing the connection of people from different cultures [6]. As a matter of fact, emojis can be useful tools to analyze social media because first, they are widely used by people from different countries and second, they have been adopted in different social media, such as Facebook, Twitter, etc. Furthermore, they are employed for purposes other than social media, such as mobile phone notification using emojis [10]. On the other hand, some emojis are ambiguous in their meaning leading to different usages. One of the most common cases is the "Person With Folded Hands" (🙏), which in some cultures (such as in Japan) is seen as "please" or "thank you", while in others (such as in Brazil) is widely used as a sign for prayer or "amem".

2 Related Works

One of the works to help computers understand emojis, attempts to build an inventory of meanings for emojis in a way that is easy for machines to understand. Wijeratne et al. [11] tried to a make connection between each emoji and its meaning in words. The output of their work is a semantic network in BabelNet. Although they try to have a comprehensive machine readable network of emojis and words, it could have been better if they considered the co-occurrence of emojis in social media with other frequent words and have an analysis on their bipartite notwork of words and emojis. Besides, a combination of emoji sentiment analysis [7] with words may give us a more accurate list of emoji meanings. In [1] a vector space model has been used for Twitter data in order to connects emojis to meaningful corresponding words.

The number of emojis that are being used in Twitter can be found on emoji-tracker.com. Furthermore, in [9], the authors discuss social aspects related to emoji usage; they argue that Twitter users who embrace emojis tend to keep using them

[1]https://www.appboy.com/blog/emojis-used-in-777-more-campaigns/.

[2]http://www.nytimes.com/2014/07/27/fashion/emoji-have-won-the-battle-of-words.html.

instead of emoticons, thus the number of emoticons being used is falling down. The study on emoji usage has also been done in a geocentric way. Scholars focused on the emoji distribution both over the world and in countries. For instance, Ljubešic and Fišer [5], gathered information about emoji usage distribution by country and investigated the emoji popularity for the whole world in this geocentric approach. Then found the list of popular emojis for each region, followed by a clustering of the countries based on emoji popularity, they found that countries could be classified into four different group based on the "most distinctive emojis". Finally, they discovered a correlation between some emojis and some world development indicators of the world bank. For example, surprisingly, countries with high life expectancy use "face with tears of joy" (😂) less often than the countries with low life expectancy.

As we mentioned before, having a network of emojis based on their co-occurrence may help us analyzing emojis from a different angle. Lu et al. [6] concentrated on trying to understand human behavior in the context of culture from data gathered from users of smartphones. Accordingly, the authors correlate the culture index with emoji sentiments. They considered the cultural index introduced by Hofstede in [4] that delineated the social differences with six features. For example, power distance is one of them. This feature expresses how much people with less power accept that power is distributed unevenly. They discovered that strong power-distance countries use more negative emotions with emojis.

3 Data Handling and Network Extraction

This is based on the subjects of tweets, we define "subjects" as the theme used for the collection of theses datasets. In this initial work, we selected two diametrically different datasets in order to verify possible structural differences. Recall that our approach argues that the structure may be linked to the subject of the conversation. We created two emoji networks, one for each dataset. The list of emojis used here are from https://apps.timwhitlock.info.

The datasets were named *WWC* and *ProgLang*. For *WWC*, the tweets were collected during the 1 mon period of the Women World Cup and Americas Cup (soccer) held in the USA in June 2015. This dataset contains more than 10 million individual tweets. The *ProgLang* dataset contains tweets from September the 20th to November the 1st in 2016 related to computer programming languages. The dataset contains approximately 2.5 million tweets.

3.1 Building an Emoji Network from Tweets

The process of creating an emoji network from tweets is quite straightforward. The general idea consists of sifting through each tweet and looking for emojis in the dataset. Each tweet generates a k-clique where k is the number of emojis in that

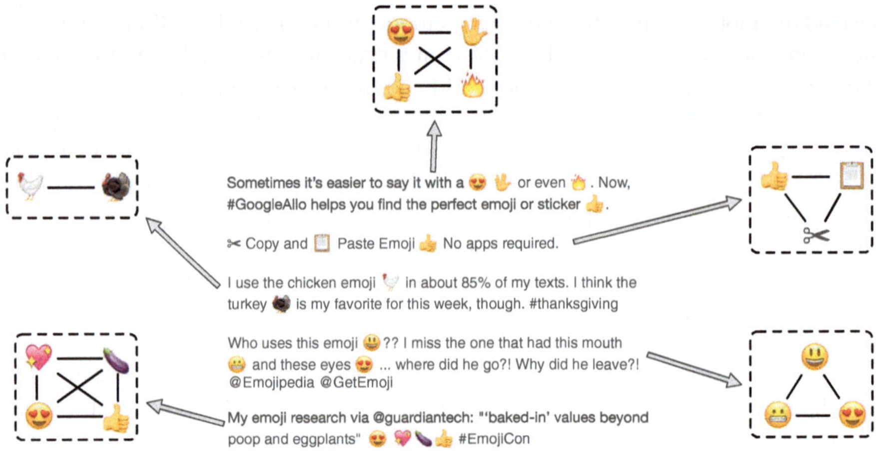

Fig. 1 Process describing the extraction of a k-clique from each individual tweets, where k is the number of emojis in each tweet. The network is undirected

Fig. 2 Result of the combination of all the k-cliques in Fig. 1. This network is undirected and weighted

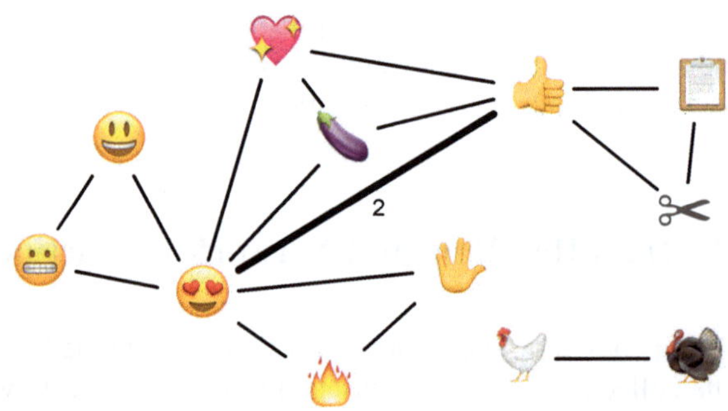

tweet; Fig. 1 shows the process in which networks are extracted from the unstructured textual data.

Following the generation of these k-cliques, they are combined forming a larger network. The resulting network for the example shown in Fig. 1 is depicted in Fig. 2. Note that the network is weighted and the edge (👍,😍) has weight 2 because it appears in Fig. 1 in two different tweets.

4 Experimental Results

After creating the networks of emojis, we performed an analysis of their structure using using Network Science concepts, such as degree distribution, edge-weight distribution, network density, to name a few [3]. We start with general network characterization but we also discuss similarities and differences between the two networks we worked on.

Table 1 Properties of the two networks used in this study. The *WWC* network is a lot more dissortative, while the programming languages is neutral

Dataset	Max weight	Average weight	Max node betweenness	Max edge betweenness	Assortativity
ProgLang	510	3.73	😁 & 🍃	(❤️,🕐)	−0.066
WWC	71,099	34.08	😁 & 🌵	(😏,🈺)	−0.193

4.1 Network Characteristics

There are several important characteristics that can be extracted from networks. Table 1 shows some basic network properties for both networks. It also shows the result of three important aspects in these networks, **Node Betweenness**: A node has high betweenness if it happens to frequently be in the shortest path between other pair of nodes. **Edge Betweenness**: An edge has high betweenness if it happens to frequently be in the shortest path between pair of nodes. **Assortativity**: It is a measure of how often a node with a particular degree connects to others of similar degree. High assortativity means that nodes connect to others alike; the metric assumes values between -1 and $+1$ for dissortativity and assortativity respectively [12].

We calculated the assortativity of the network; both networks are slightly disassortative meaning that the nodes with higher degree tends to have connections with nodes with lower degree. The *WWC* network is more dissortative.

For the analysis of betweenness, Table 1 shows the grinning face (😁) has the highest node betweenness in both datasets confirming the popularity of this emoji regardless of the subject area. Another interesting result from Table 1 is the fact that the maximum edge betweeeness occurs for the edge linking the smirking face (😏) and squared Chinese-Japanese-Korean character (🈺). It is amusing because smirking face is one of the top favorite emojis in the United States [6], the squared cjk is related to Japanese characters, and the final of the 2015 world cup was USA against Japan. It appears then that if one knows the semantics of these emojis, it may be possible to learn something about the subject area from which they were extracted and this indications opens a door for possible recommendation systems.

4.2 Degree and Weight Degree Distributions

One of the most common characteristics scientists measure in a weighted network are both the degree and weighted degree distributions of nodes [2]. We tried to fit common functions found in real-world networks and used log-likelihood ratio—denoted by $L(d_1, d_2)$—for distribution analysis. The positive values of log-likelihood tell us that the left function d_1 is a better fit to the original data, and d_2 otherwise.

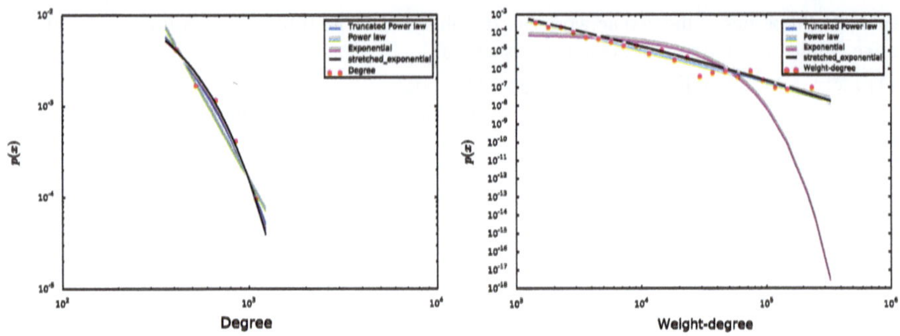

Fig. 3 Fitting applied to the degree and weighted-degree distributions for *WWC*

Table 2 Log-likelihood ratio ($log L$) for degree, weighted degree and edge-weight distributions for the *WWC* network

Functions	Degree		Weighted degree		Edge weight	
(d_1, d_2)	$log L$	p-value	$log L$	p-value	$log L$	p-value
(powerlaw, exponential)	−3.96	0.015	173.35	0.000	878.31	0.000
(powerlaw, truncated power-law)	−3.59	0.007	−2.77	0.018	−5.01	0.001
(powerlaw, stretched exponential)	−4.25	0.063	−1.98	0.243	−2.44	0.544
(truncated power-law, exponential)	−0.37	0.289	176.13	0.000	883.32	0.000
(truncated power-law, stretched exponential)	−0.66	0.509	0.79	0.263	2.57	0.279
(exponential, stretched exponential)	−0.29	0.444	−175.34	0.000	−880.74	0.000

In Fig. 3 we demonstrate several possible fitted functions for degree and weighted degree distributions of the *WWC* data set. The red dotted line show the data and the lines are the functions that could possibly fit the data distribution. A visual inspection can immediately say that the exponential function is not a good fit for weight-degree distribution. For a more complete analysis of the goodness of fit, we show in Table 2 the log-likelihood ratio and the p-value between different functions with respect to *WWC* data set. The results show us that stretched exponential is the best fit function for degrees.

In addition to the degree analysis, another important aspect of our emoji network are the weights of edges. The edge weight represents how pronounced the co-occurrence of pairs of emojis are in the dataset. Hence it is important to characterize this distribution to understand how the values of edges are distributed. Figure 4 shows the fitting of the edge-weight distribution for *WWC* network. We also preformed a log-likelihood analysis and found that the best fitted distribution of this is a truncated

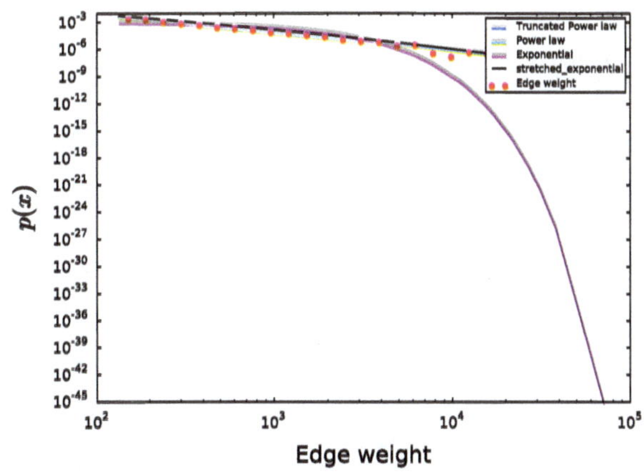

Fig. 4 Edge-weight distribution for the *WWC* Network

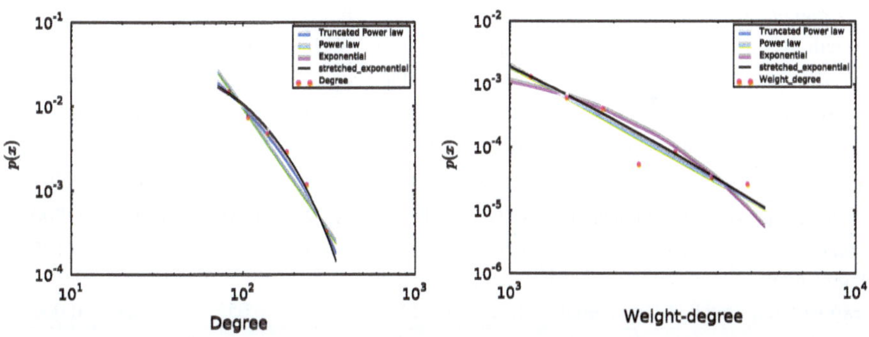

Fig. 5 Fitting applied to the degree and weighted-degree distributions for the *ProgLang* network

power-law. This means that there are relatively fewer pair of emojis that are popular and that most pairs are rare.

In this paper we also reconstructed an emoji network from another dataset related to programming languages. Similar to what we have done for the *WWC* network, we analyzed the network degree and weighted degree distributions, as well as the edge-weight distribution. The degree distributions are depicted in Fig. 5.

Furthermore, Fig. 6 shows the best fitted function for the edge-weight distribution as being a truncated power-law which again agrees with what was found for the *WWC* network.

The fitting of the functions was done again using an approach based on the log-likelihood ratio. In Table 3 we find more details about the pairwise comparison between different functions for degree, weighted-degree, and edge-weight. As one can observe, the best fitted function favors a stretched exponential for degrees, while for weighted-degree and edge-weight, the truncated power law is clearly the best fit.

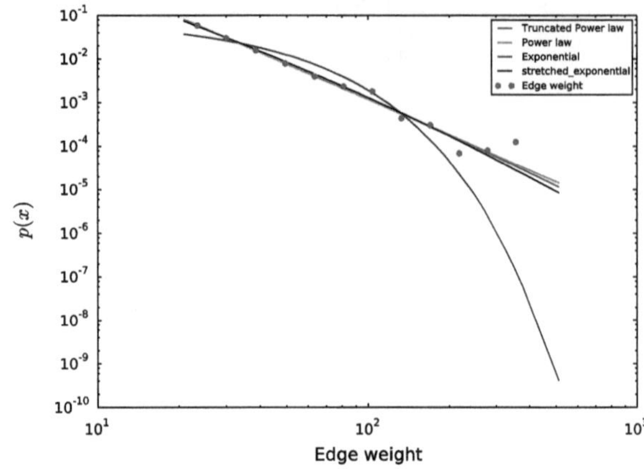

Fig. 6 Edge-weight distribution for the *ProgLang* Network

Table 3 Log-likelihood ratio ($log L$) for degree, weighted degree and edge-weight distribution for the *ProgLang* network

Functions	Degree		Weighted degree		Edge weight	
(d_1, d_2)	$log L$	p-value	$log L$	p-value	$log L$	p-value
(powerlaw, exponential)	−11.08	0.002	1.72	0.347	100.04	0.000
(powerlaw, truncated power-law)	−10.30	0.000	−0.17	0.556	−0.52	0.310
(powerlaw, stretched exponential)	−11.08	0.002	−0.09	0.782	1.54	0.397
(truncated power-law, exponential)	−0.77	0.499	1.92	0.215	100.56	0.000
(truncated power-law, stretched exponential)	−0.78	0.525	0.08	0.206	2.07	0.105
(exponential, stretched exponential)	0.01	0.922	−1.84	0.055	−98.48	0.000

In summary, in both data sets, we have the same type of distribution for degree, weighted-degree and edge-weight values of the networks. This is a preliminary work but it does seem to indicate that structure of emoji usage is not much affected by the subject of the conversation. Note that this does not mean that the emojis used are the same, quite the contrary, our work only argues that the networks formed from the co-occurrence have similar structures but it is very likely that different emojis occupy similar structural positions in the different networks. Table 1 supports this claim in our two datasets.

5 Conclusion

In this paper we constructed co-occurrence networks from emojis and analyzed their structure to understand possible regularities. We used two datasets and showed that although they do not seem to have a structure similar to network of words in written language or other common real-world networks, they do have similar structures among the two datasets.

We are working on larger datasets. In these, we will focus on community detection as a way to find family of emojis and whether the families correlate to classes of emoji (flags, professions, etc.) Furthermore, PageRank could be useful to understand the importance of emojis to language; for this we need to have a directed version and we are investigating if the order they appear in the text could realistically represent a direction. For instance if one writes “I ❤️ to have a 🍺” or something such as “🍺 is one of the things I ❤️” have slightly different meanings due to the order the emojis are used but also the relation to the words in the sentence. A directed network of usage could capture some of these nuances.

References

1. Barbieri, F., Ronzano, F., Saggion, H.: What does this emoji mean? a vector space skip-gram model for twitter emojis. In: Language Resources and Evaluation conference, LREC, Portoroz, Slovenia (2016)
2. Barrat, A., Barthelemy, M., Pastor-Satorras, R., Vespignani, A.: The architecture of complex weighted networks. Proc. Natl. Acad. Sci. U.S.A. **101**(11), 3747–3752 (2004)
3. Borgatti, S.P., Mehra, A., Brass, D.J., Labianca, G.: Network analysis in the social sciences. Science **323**(5916), 892–895 (2009)
4. Hofstede, G., Hofstede, G.J., Minkov, M.: Cultures and Organizations: Software of the Mind, vol. 2. Citeseer (1991)
5. Ljubešic, N., Fišer, D.: A global analysis of emoji usage. In: ACL 2016, pp. 82 (2016)
6. Lu, X., Ai, W., Liu, X., Li, Q., Wang, N., Huang, G., Mei, Q.: Learning from the ubiquitous language: an empirical analysis of emoji usage of smartphone users. In: Proceedings of the 2016 ACM International Joint Conference on Pervasive and Ubiquitous Computing, pp. 770–780. ACM (2016)
7. Novak, P.K., Smailović, J., Sluban, B., Mozetič, I.: Sentiment of emojis. PloS one **10**(12), e0144296 (2015)
8. Panksepp, J.: Affective consciousness: core emotional feelings in animals and humans. Conscious. Cogn. **14**(1), 30–80 (2005)
9. Pavalanathan, U., Eisenstein, J.: Emoticons vs. emojis on twitter: a causal inference approach (2015). arXiv:1510.08480
10. Tauch, C., Kanjo, E.: The roles of emojis in mobile phone notifications. In: Proceedings of the 2016 ACM International Joint Conference on Pervasive and Ubiquitous Computing: Adjunct, pp. 1560–1565. ACM (2016)
11. Wijeratne, S., Balasuriya, L., Sheth, A., Doran, D.: Emojinet: building a machine readable sense inventory for emoji. In: International Conference on Social Informatics, pp. 527–541. Springer (2016)
12. Xulvi-Brunet, R., Sokolov, I.M.: Changing correlations in networks: assortativity and dissortativity. Acta Phys. Pol. B **36**, 1431 (2005)

language on other common real-world networks, they do have similar structure among the two datasets.

Characterization of Written Languages Using Structural Features from Common Corpora

Younis Al Rozz, Harith Hamoodat and Ronaldo Menezes

Abstract For more than 5,000 years, we have been communicating using some form of written language. For many scholars, the advent of written language contributed to the development of societies because it enabled knowledge to be passed to future generations without considerable loss of information or ambiguity. Today, it is estimated that we use about 7,000 languages to communicate, but the majority of these do not have a written form; in fact, there are no reliable estimates of how many written languages exist today. There are three main families of written languages: Afro-Asiatic, Indo-European, and Turkic. These families of languages are based on historical family-trees. However, with the amount of data available today, one can start looking at language classification using regularities extracted from corpora of text. This paper focus on regularities of 10 languages from the mentioned families. In order to find features for these languages we use (*1*) Heaps' law, which models the number of distinct words in a corpus as a function of the total number of words in the same corpora, and (*2*) structural properties of networks created from word co-occurrence in large corpora for different languages. Using clustering approaches we show that despite differences from years of being used in separate countries, the clustering still seem to respect some historical organization of families.

Keywords Co-occurrence networks · Language classification · Heaps' law · Clustering

Y. Al Rozz (✉) · H. Hamoodat · R. Menezes
BioComplex Laboratory, School of Computing Florida Institute of Technology, Melbourne, USA
e-mail: yyounis2013@my.fit.edu

H. Hamoodat
e-mail: hhamdon2013@my.fit.edu

R. Menezes
e-mail: rmenezes@cs.fit.edu

B. Gonçalves et al. (eds.), *Complex Networks VIII*,
Springer Proceedings in Complexity, DOI 10.1007/978-3-319-54241-6_14

1 Introduction

The development of society cannot be said to be caused by the advent of writing but writing is certainly linked to modern life as it only appeared around 5,000 years ago. According to Coulmas [15], writing is the most important "sign system" ever invented. It is quite difficult to imagine our society thriving without books, research articles, instruction manuals, lecture notes, etc. The importance of writing is even recognized by many cultures and often its invention is attributed to divine intervention such as god Ganesh in India, or the god Thoth in ancient Egypt.

Writing enables the transmission of information between many generations without any loss of information; it broadens the range of communication of individuals. Today, it is estimated that humans use about 7,000 languages to communicate,[1] although this number is in decline as languages become extinct. Moreover, the majority of these do not have a written form; in fact, there are no reliable estimates of how many written languages exist today. Linguists have been studying languages and how they should be organized for a long time [11], however most classifications are based on historical or phonetic approaches. There are many families of languages, and few are well known such as: Uralic, Afro-Asiatic, Indo-European, and Turkic. Figure 1 shows a sample of the Indo-European set of languages.

The advances in Network Science and Natural Language Processing (NLP) in recent years has motivated researchers to utilize both disciplines together in language classification.

Nowadays studies can be done quantitatively and not only qualitatively. It is quite common to have data regarding any subject of interest. In the context of text analysis, the studies range from discovering language structure [30], classification of languages into families [6, 19, 23, 24], word tagging problems [10], machine translation [2], summarization systems [3], to the improvement of search engines and information retrieval (IR) [28]. Although we review a few of the related work in Sect. 2, an interested reader can find a deeper analysis of the literature in [30].

The understanding of structural language similarities can lead to metrics to evaluate the quality of one's writing, translations, and even classification of literary styles. It is quite possible that different styles present different writing structures. In this work, we show that even without semantic analysis of the text itself, and focusing solely on the structure built from syntax, we can reveal that characteristics of many languages are common. More specifically, we used statistical measures of a word co-occurrence networks as well as regularities extracted from parameters of Heaps' law to classify 10 world languages. The classification process was performed using two methods: K-Means, and Hierarchical Clustering.

[1] https://www.ethnologue.com.

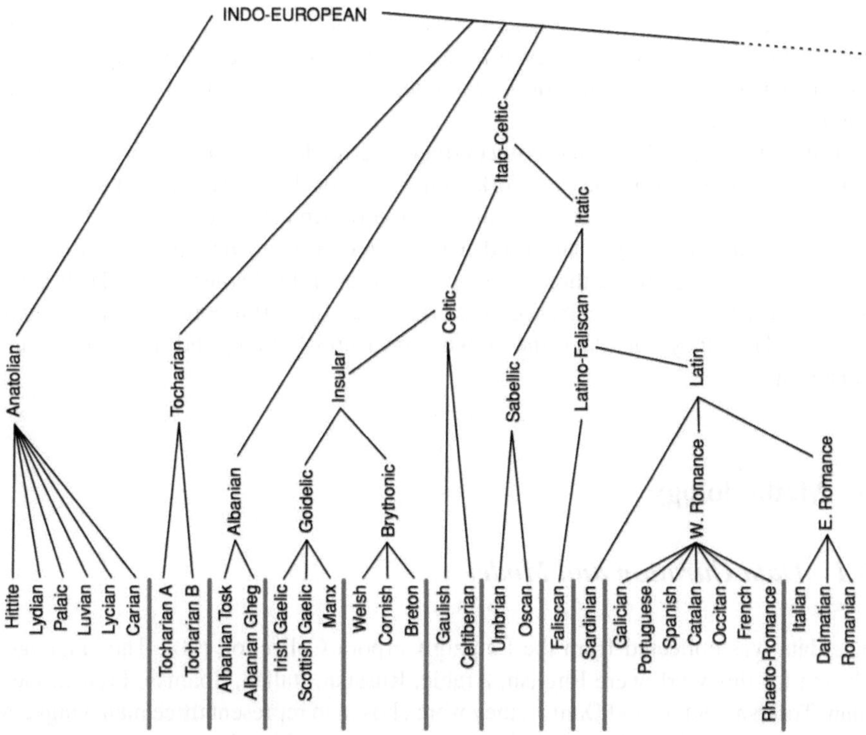

Fig. 1 Part of the family tree of Indo-European languages (adapted from [11])

2 Related Work

Many researchers have investigated the possibility of using statistical and mathematical modeling to understand regularities in written languages. Choudhury and Mukherjee [13] discuss many ways in which networks can be created from text but they all fall into two main categories: lexical networks and word co-occurrence networks. The first category is concerned with cognitive systems and Psycholinguistics studies [7] and can be further classified into phonological [4], semantic [32], and orthographic networks [14]. Phonological networks can be a network of phonemes [27] or syllables [29]. The second type of language network can be further categorized into co-location [25] and syntactic dependency networks [22].

The attempt to use language structure as a classification tool is not entirely new. In fact, Song [31] discussed the concept of *linguistic typology* as a field which looks at the comparison of languages (search for similarities and differences) across all levels of language structure such as syntax, semantics, morphology, and phonology. Three types of linguistic typology exist [8]: qualitative, quantitative, and theoretical.

Liu and Xu constructed syntactic networks for 15 languages using word and lemma form. They analyzed seven network parameters to classify languages and found that word-formed networks are better than lemma networks in classifying languages [24].

Liu and Cong [23] created co-occurrence networks from a text in 14 different languages and used complex network parameters for their classification using hierarchical clustering. Ban et al. [6] built a co-occurrence network using text from five books for three languages and used network measures to find the similarity and differences between those three languages. Gao et al. [19] constructed six directed and weighted word co-occurrence networks based on 100 reports from the United Nations. Then they compared the network measures but they did not perform any clustering.

3 Methodology

3.1 Data Curation and Model

The data was collected from the Leipzig Corpora Collection [20]. The languages chosen for this work were English, Arabic, Russian, Italian, Spanish, French, German, Turkish, Dutch, and Danish; they were chosen to represent three main language families, namely Afro-Asiatic, Indo-European, and Turkic. The text corpus for each language was constructed from Wikipedia and news pages to ensure some vocabulary diversity and a good representation for each language. The size of the corpus for each language is consistently made of one million sentences. The entire text was converted to lower case, then punctuation and special characters were removed. This work looks at language structure for meaningful words and sequences; stop words (e.g. prepositions, articles, etc.) were removed from the text. These so-called functional words can skew the statistical representation of the words in particular in the context of network science (described later).

3.2 Feature Extraction

One of the best-known characteristics of vocabulary is the Heaps' law (also known as Herdan's law) introduced in the 1960s [21] which describes the vocabulary growth in texts [18]. The law is defined as:

$$V_R(n) = Kn^{\beta}, \tag{1}$$

where V_R is the number of vocabulary words in the text of size n, and K and β are parameters determined experimentally.

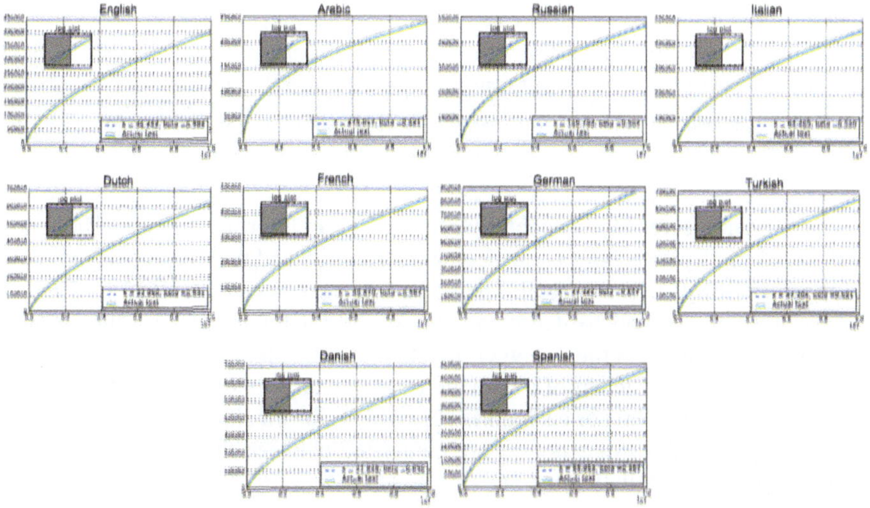

Fig. 2 Fitting of Heaps' law for the 10 languages used in this study (and the value of K and β respectively)

Table 1 From top to bottom and from left to right the languages in Fig. 2. The values of K and β from Eq. 1 is shown

	English	Arabic	Russian	Italian	Dutch	French	German	Turkish	Danish	Spanish
K	34.24	275.01	146.79	64.46	23.95	49.87	27.32	57.30	21.84	55.95
β	0.58	0.42	0.50	0.55	0.63	0.56	0.64	0.58	0.63	0.55

Heaps' law represents the vocabulary richness of a certain language, a large text corpus of 10 million words was used for the fitting of the Heaps' law parameters Fig. 2. These parameters are used as a part of the features vector that will be used to characterize the 10 languages used in this work.

Table 1 shows the values of K and β for the fitting in Fig. 2. For English, the values of K are expected to be between 10 and 100 and the values β between 0.4 and 0.6. Our results agree with this expectation but the values of K for Arabic and Russian is greater than 100.

After the fitting of Heaps' law to our corpora, we set to create co-occurrence word networks. Our networks are simple and link words in each corpus if they are adjacent to each other in text. Hence, nodes represent unique words and edges represent the connection between each two consecutive words. The edges' weights represent the frequency in which the two words appear next to each other. Table 2 shows the size of each network in terms of number of nodes n and number of edges m.

The generation of the networks gives us the structure and the values for n and m. Note however from Table 2 that for all languages the values of n and m are very similar which indicates they are not good features to let us characterize the languages. However, there are other structural characteristics that can be computed from the networks.

Table 2 Size of the word co-occurrence networks for all 10 languages

	English	Arabic	Russian	Italian	Dutch	French	German	Turkish	Danish	Spanish
n	18,986	29,995	37,341	31,361	30,475	30,248	39,098	34,945	30,329	29,999
m	77,989	81,046	93,587	94,494	94,427	94,611	95,774	89,385	88,985	94,919

The average degree $\langle k \rangle$ is generally provided as an information item. These networks tend to display a power-law degree distribution and the average degree does not represent the distribution well. The highest average degree was 8.21 for English and the lowest was 4.89 for German. The reason for this is because the German language's vocabulary is much bigger than that of English [9].

The clustering coefficient of a network (C) is given by the average clustering of the clustering coefficients of each node (C_i) which (informally) captures the extend to which the neighbors of a node i are connected between themselve, this can be calculated using the equation below:

$$C_i = \frac{2E_i}{k_i(k_i - 1)}, \tag{2}$$

where, E_i is the number of links that exist between the neighbors of node i, and the denominator number of possible links that could exist between nodes i.

Russian and Arabic have the lowest clustering coefficient: 0.012 and 0.019 respectively; English and Danish score the highest: 0.047 and 0.041 respectively. This is due to the fact that Russian and Arabic are morphological languages, which means that they have more word forms than analytic languages such as English and Danish [1].

Another vital characteristic for networks analysis is the average path length. We know that social networks have high C and low average path length (ℓ) computed as:

$$\ell = \frac{1}{N(N-1)} \sum_{i \neq j} d_{ij}, \tag{3}$$

where d_{ij} is the distance between nodes i and j. Russian has the longest value for ℓ with 4.91 steps, while the shortest one was 3.82 for English. Again, this happens because morphological languages like the Russian and Arabic tend to have a longer path than analytic languages like English and Dutch [1].

Networks can be divided into consistent groups of nodes called communities [16] whose density of edges within the community is higher than outside it. There are many algorithms in the literature proposed to find these communities but one of the classical ways is to calculate the modularity of the network (Q). We computed the value of Q for all 10 networks using the approach proposed by Newman [26]. Based on this metrics, Russian has the largest modularity value of 0.481, while the lowest value was 0.379 scored by English.

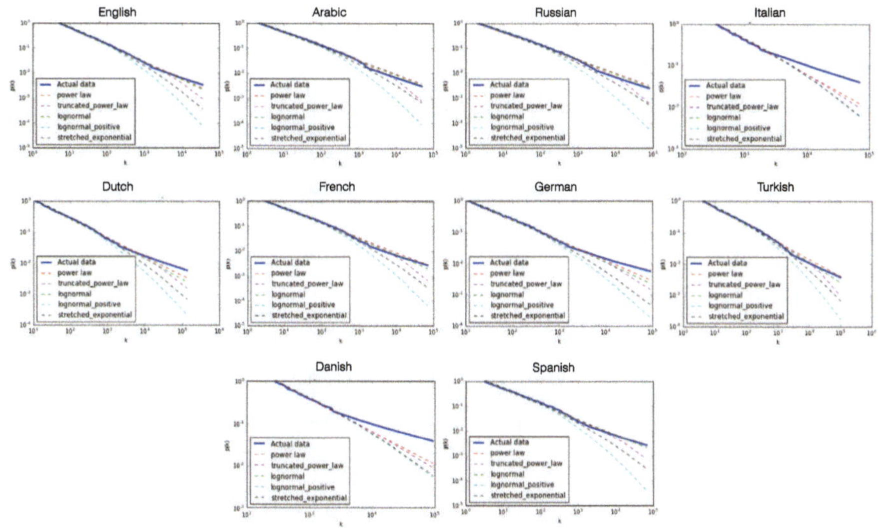

Fig. 3 Fitting of the degree distribution

The last two parameters, α_d and α_s were obtained by fitting functions to weighted degree distribution of the network and size distribution of communities of words. As shown later in Table 3, the values of α_d are quite close to what is expected for real-world networks ($2 \leq \alpha \leq 3$). We believe the reason for the lower exponent values was the removal of the functional words. Figure 3 shows that a power law function (i.e. $P(k) \sim k^{\alpha}$, where k represents the node degrees) has the best fit when compared to other common functions of real-world networks.

Similarly, the α_s value for the distribution of community size shows a good fit with a power law function, which is expected also in real-world networks with community structure; according to Arenas et al. [5] the distribution of community sizes in real network appear to have a power law form $P(s) \sim s^{\alpha}$. Both exponents have been used as part of the feature vector representing the languages. Figure 4 shows the fitting for the community size for all 10 languages and Table 3 shows the values for α_s.

For each of the networks we built, we generated random networks with the same size and using the Erdös-Rényi model. The purpose was to analyze the clustering of our word networks in comparison with a random network. The average clustering coefficient values for the random networks were much smaller than those in the word networks. For example, in Italian, the average clustering coefficient for our network is 0.022 while in the random network was 0.00019. Also, the average path length (ℓ) for the 10 languages was between 3.8 and 4.9 which means our networks appear to be small-world [33].

After all the analysis we had an 8-dimension feature vector for each language as depicted in Table 3. In the next section, we will use these vectors to do a clustering of the languages leading to a classification of them based on their structural similarities.

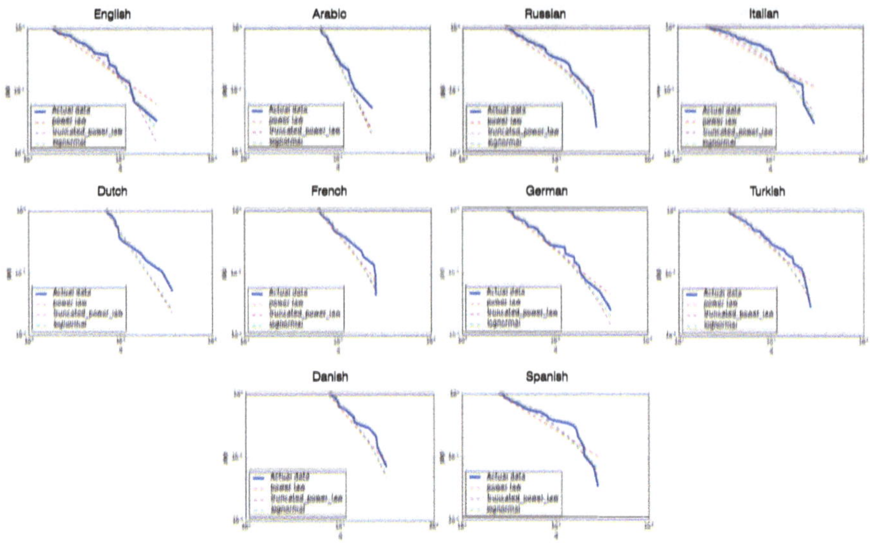

Fig. 4 Fitting of the size distribution in the power law package

Table 3 Each line in this table represent 8-dimension feature vector for the language shown in the first column

Languages	β	K	$\langle k \rangle$	C	ℓ	Q	α_d	α_s
English	0.582	34.242	8.215	0.047	3.824	0.379	1.827	2.070
Arabic	0.421	275.017	5.404	0.019	4.454	0.466	1.508	3.937
Russian	0.502	146.793	5.012	0.012	4.910	0.481	1.660	2.037
Italian	0.550	64.465	6.026	0.022	4.280	0.405	1.751	1.800
Dutch	0.631	23.950	6.197	0.026	4.194	0.388	1.725	3.186
French	0.567	49.879	6.255	0.023	4.213	0.385	1.745	2.774
German	0.647	27.322	4.899	0.023	4.471	0.464	1.689	2.194
Turkish	0.581	57.304	5.115	0.023	4.430	0.471	1.716	2.223
Danish	0.636	21.849	5.868	0.041	4.200	0.438	1.740	2.761
Spanish	0.557	55.955	6.328	0.023	4.239	0.389	1.730	1.934

4 Results and Discussion

We have executed clustering using two known algorithms: K-Means and Hierarchical Clustering. Recall that the purpose of this work is to classify languages according to the features extracted from Heaps' law and network properties.

4.1 K-Means Clustering

K-Means is a fast and widely-used clustering algorithm that works by minimizing the sum-of-squares distance of the data points within the cluster. The number of clusters must be specified in advance, so two methods were used to find the optimal number of clusters. The first one is the silhouette method; it provides a visual aid in determining the number of clusters. The silhouette coefficient which ranged between -1 and 1 indicates the closeness of each data point in a cluster to other points in the neighboring clusters. After that, we used the elbow method to validate the number of clusters found in the silhouette method.

Due the high dimensionality of the feature vectors, we run a Principle Component Analysis (PCA) to reduce the dimensionality of the features vector to two dimensions so that the resulting K-Means clusters can be visualized. We also wanted to independently check whether the parameters extracted from the Heaps' law were providing extra information to the clustering of the feature vectors. The silhouette method was applied with and without the two Heaps' law parameters (K and β). In the first case, the optimal number of clusters was three. When the Heaps' parameters were added, the silhouette plot suggests a number of clusters between four and five as a good choice (Fig. 5). These results indicate the importance of the Heaps' parameters to the process of the language classification.

The elbow method was used to validate the optimal number of clusters found by the silhouette method. The elbow plot suggests an optimal number of three clusters when the two Heaps' parameters are not considered, which agreed with the results of the silhouette method. The result of the K-means clustering for this case was

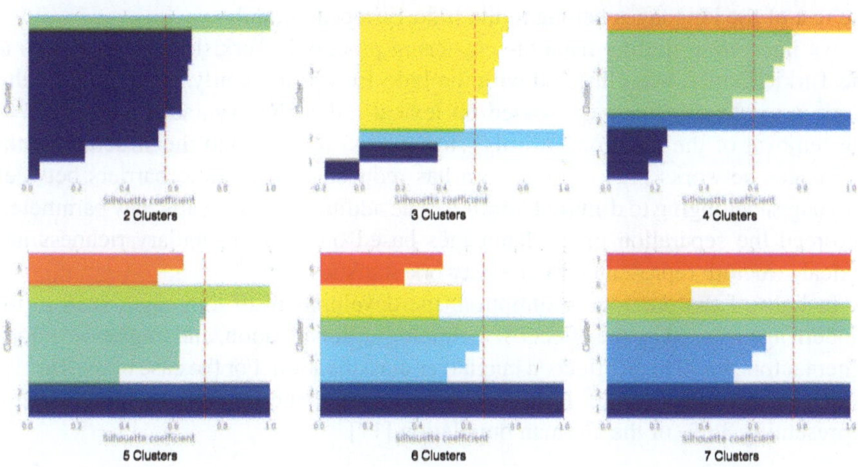

Fig. 5 Silhouette analysis on K-Means clustering where the value of the Heaps' law parameters were included after the PCA. The same analysis has been done for the case without the use of the Heaps' law parameters which we did not include a picture due to space limitations in the paper

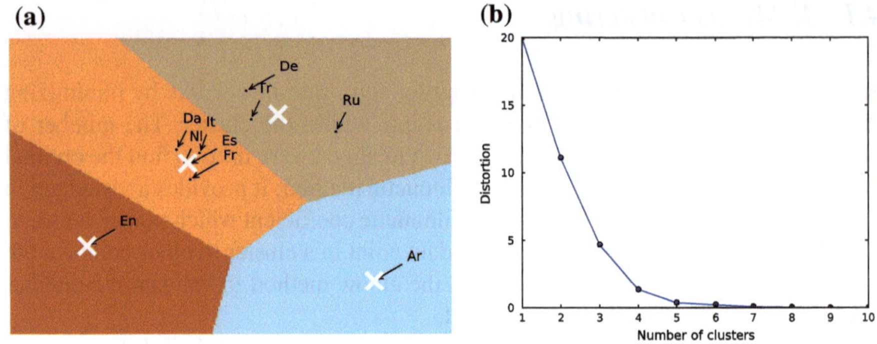

Fig. 6 **a** K-means clustering after PCA and using Heaps' law parameters and network parameters. **b** The elbow rule shows that four clusters appear to be the best choice for the K-means.

that Italian, Spanish, German, Russian, and Turkish clustered together. The second cluster contains French, Danish, Dutch, and English, while Arabic appeared in its own cluster.[2] When adding the parameters of the Heaps' law, the elbow of the curve indicates an optimal number of four clusters (Fig. 6b). In this case, Italian, Spanish, French, Danish, and Dutch were clustered together. The second cluster contains Russian, German, and Turkish, while English and Arabic separated into their own clusters (Fig. 6a), which also supports the results of the silhouette method indicating the importance of Heaps' parameters to the classification process and the fact that the complete set of parameters offers a higher granularity for the clustering. These results match, to a certain degree, the linguistic typology classification of languages into genetic families as the Arabic language belongs to the Afro-Asiatic family, while the rest of the languages belong to the Indo-European Family.

An interesting finding from the clustering process is Turkish, which belongs to the Turkic family, was clustered with the Indo-European Family. As the aim of this work is to classify languages based on lexical rather than syntactical perspective, the removal of the functional words (stop words) has affected the structure of the languages networks [12]. This in turn has reduced the syntactic barriers between languages belonging to different families. The addition of the Heaps' law parameters enforced the separation of the languages based on their vocabulary richness and lexical structure represented by the network statistics.

In light of the previous assumption, the development of languages seen in the modern age, caused by the effects of technology, globalization, and migration among other factors, has had on effect on languages classification. For the case of the Turkish language, as of the year 2011, three million Turkish people were living in Germany, representing 3.6% of the German population [17].

---

[2]We again decided to show the charts only for the case with the Heaps' parameters due to space restrictions.

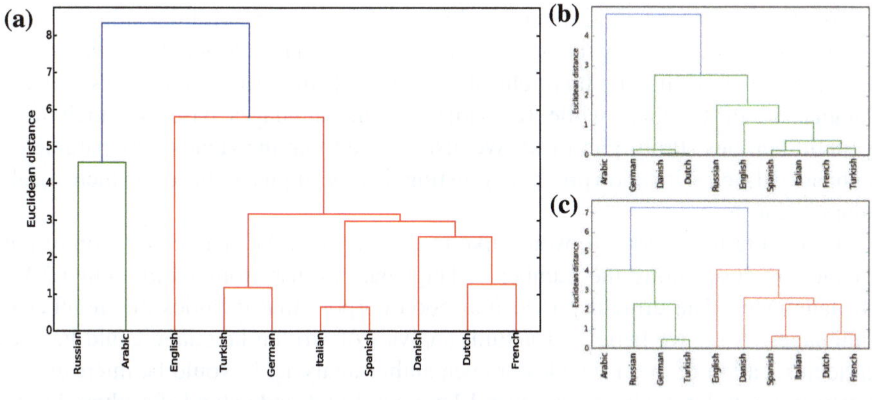

Fig. 7 Hierarchical clustering of the 10 languages used in our study. **a** Shows the classification using the network parameters as well as the Heaps' law parameters while **b** shows the classification using Heaps' law parameters and **c** network parameters separately

4.2 Hierarchical Clustering

The results of K-means clustering can only classify languages from the top level of the family tree. To find the relationships between languages in a more structured way we applied a hierarchical clustering to the language feature vectors. In this case, we decided to also test whether the Heaps' law features alone would provide a similar classification to the classification based on network features alone. Figure 7b show the classification using only the Heaps' parameters while Fig. 7c shows the same results using only network parameters. Although both classifications have interesting characteristics that resemble traditional language classifications, the combination of both features in Fig. 7a yields a classification that appears to be enhanced. For instance, the distance between the Turkish and German languages was increased.

5 Conclusion

The understanding of languages and their characterization has again become a topic of interest for the scientific community. Studies using large amounts of data may be able to provide a different view of how languages relate to one another and see possible trends or influences of one over the other.

In our study, we look at the possibility of characterizing written language solely from the point of view of structural features. We concentrated on two class of features: Heaps' law, which looks at richness of vocabulary in a language, and Network Science features extracted from the construction of word co-occurrence networks. In the process of extracting network features, we also demonstrated that these networks exhibit both scale-free and small-world properties.

We used K-Means and Hierarchical Clustering together with the silhouette and elbow methods to identify the optimal number of language clusters to the dataset we have. We showed that the hierarchical clustering distinguish relationships between languages sub-families, while K-Means clusters languages based on their main genetic families (Proto-Families). We also showed that the Heaps' law parameters enhanced the classification process by distinguishing languages based on their vocabulary richness.

Following this work, we would like to go deeper in the characterization of languages by augmenting the number of languages we use from 10 to around 30 or 40 languages. The difficulty is to find good corpora that includes this number of languages. Also, we believe structural analysis of written language could be used in identification of literary styles or even author analysis. It would be interesting to perform a similar analysis for several languages and understand if authors have a structural fingerprint in their writing style that can be identified and whether this fingerprint resist the translations of their texts.

References

1. Abramov, O., Mehler, A.: Automatic language classification by means of syntactic dependency networks. J. Quant. Linguist. **18**(4), 291–336 (2011)
2. Amancio, D.R., Antiqueira, L., Pardo, T.A.S., da F. Costa, L., Oliveira Jr., O.N., Nunes, M.G.V.: Complex networks analysis of manual and machine translations. Int. J. Mod. Phys. C **19**(04), 583–598 (2008)
3. Antiqueira, L., Oliveira, O.N., da Fontoura Costa, L., das Graças Volpe Nunes, M.: A complex network approach to text summarization. Inf. Sci. **179**(5), 584–599 (2009)
4. Arbesman, S., Strogatz, S.H., Vitevitch, M.S.: The structure of phonological networks across multiple languages. Int. J. Bifurc. Chaos **20**(03), 679–685 (2010)
5. Arenas, A., Danon, L., Diaz-Guilera, A., Gleiser, P.M., Guimera, R.: Community analysis in social networks. Eur. Phys. J. B Condens. Matter Complex Syst. **38**(2), 373–380 (2004)
6. Ban, K., Meštrović, A., Martinčić-ipšić, A.: Initial comparison of linguistic networks measures for parallel texts. In: 5th International Conference on Information Technologies and Information Society (ITIS), 97104. Citeseer (2013)
7. Beckage, N.M., Colunga, E.: Language networks as models of cognition: understanding cognition through language. In: Towards a Theoretical Framework for Analyzing Complex Linguistic Networks, pp. 3–28. Springer (2016)
8. Bickel, B.: Typology in the 21st century: major current developments. Linguist. Typol. **11**(1), 239–251 (2007)
9. Biemann, C., Bordag, S., Heyer, G., Quasthoff, U., Wolff, C.: Language-independent methods for compiling monolingual lexical data. In: International Conference on Intelligent Text Processing and Computational Linguistics, pp. 217–228. Springer (2004)
10. Brill, E.: Transformation-based error-driven learning and natural language processing: a case study in part-of-speech tagging. Comput. Linguist. **21**(4), 543–565 (1995)
11. Campbell, L., Poser, W.J.: Language Classification: History and Method. Cambridge (2008)
12. Chen, X., Liu, H.: Function nodes in Chinese syntactic networks. In: Towards a Theoretical Framework for Analyzing Complex Linguistic Networks, pp. 187–201. Springer (2016)
13. Choudhury, M., Mukherjee, A.: The structure and dynamics of linguistic networks. In: Dynamics on and of Complex Networks, pp. 145–166. Springer (2009)

14. Choudhury, M., Thomas, M., Mukherjee, A., Basu, A., Ganguly, N.: How difficult is it to develop a perfect spell-checker? A cross-linguistic analysis through complex network approach. In: TextGraphs-2: Graph-Based Algorithms for Natural Language Processing, p. 81 (2007)
15. Coulmas, F.: The Writing Systems of the World. B. Blackwell (1989)
16. de Arruda, H.F., da F. Costa, L., Amancio, D.R.: Topic segmentation via community detection in complex networks. Chaos: an interdisciplinary. J. Nonlinear Sci. **26**(6), 063120 (2016)
17. Deutschland and Statistisches Bundesamt Deutschland. *Statistisches Jahrbuch Deutschland und Internationales*. Statistisches Bundesamt (2012)
18. Font-Clos, F., Boleda, G., Corral, Á.: A scaling law beyond Zipf's law and its relation to Heaps' law. New J. Phys. **15**(9), 093033 (2013)
19. Gao, Y., Liang, W., Shi, Y., Huang, Q.: Comparison of directed and weighted co-occurrence networks of six languages. Phys. A. Stat. Mech. Appl. **393**, 579–589 (2014)
20. Goldhahn, D., Eckart, T., Quasthoff, U.: Building large monolingual dictionaries at the Leipzig corpora collection: from 100 to 200 languages. In: *LREC*, pp. 759–765 (2012)
21. Herdan, G.: Type-Token Mathematics, vol. 4. Mouton (1960)
22. i Cancho, R.F.: The structure of syntactic dependency networks: insights from recent advances in network theory. In: Problems of Quantitative Linguistics, pp. 60–75 (2005)
23. Liu, H.T., Cong, J.: Language clustering with word co-occurrence networks based on parallel texts. Chin. Sci. Bull. **58**(10), 1139–1144 (2013)
24. Liu, H., Chunshan, X.: Can syntactic networks indicate morphological complexity of a language? EPL (Europhys. Lett.) **93**(2), 28005 (2011)
25. Mamede, N., Correia, J., Baptista, J.: Syntax deep explorer. In: Computational Processing of the Portuguese Language: 12th International Conference, PROPOR 2016, Tomar, Portugal, July 13–15, 2016, Proceedings, vol. 9727, p. 189. Springer (2016)
26. Newman, M.E.J.: Modularity and community structure in networks. Proc. Natl. Acad. Sci. **103**(23), 8577–8582 (2006)
27. Siew, C.S.Q.: Community structure in the phonological network. Front. Psychol. **4**, 553 (2013)
28. Singhal, A.: Modern information retrieval: a brief overview. IEEE Data Eng. Bull. **24**(4), 35–43 (2001)
29. Soares, M.M., Corso, G., Lucena, L.S.: The network of syllables in Portuguese. Phys. A Stat. Mech. Appl. **355**(2), 678–684 (2005)
30. Solé, R.V., Corominas-Murtra, B., Valverde, S., Steels, L.: Language networks: their structure, function, and evolution. Complexity **15**(6), 20–26 (2010)
31. Song, J.J.: The Oxford Handbook of Linguistic Typology. Oxford University Press (2010)
32. Steyvers, M., Tenenbaum, J.B.: The large-scale structure of semantic networks: Statistical analyses and a model of semantic growth. Cogn. Sci. **29**(1), 41–78 (2005)
33. Watts, D.J., Strogatz, S.H.: Collective dynamics of small-worldnetworks. Nature **393**(6684), 440–442 (1998)

Optimal Information Security Investment in Modern Social Networking

Andrey Trufanov, Nikolay Kinash, Alexei Tikhomirov, Olga Berestneva and Alessandra Rossodivita

Abstract For further clarification of methodological issues of the social network's information security we stratified the systems that support human relations into three components of different nature: computer, communication and social ones. A security model for a network component is developed using consideration of security for individual nodes. Modeling of attacks on networks in whole is analyzed taking into account specification of network security level. The results for real computer, communication and social entities supported that for a network attacked intentionally it is better off allocating the investment proportionally to degree centralities of the nodes rather than uniformely. The analysis further hints that to make investment justifiable to protect a network, its proprietor should spend lesser than to reach approximately 0.4 of network security level.

1 Introduction

Practical social networking has been observed almost as long as societies themselves have existed, but the fantastic possibilities of modern technologies made a new scope on the issue to recognize and exploit pertinent links comprehensively [1] and thus to

A. Trufanov (✉) · N. Kinash
Irkutsk National Research Technical University, Irkutsk, Russia
e-mail: troufan@istu.edu

N. Kinash
e-mail: kinash_family@mail.ru

A. Tikhomirov
Inha University, Incheon, Republic of Korea
e-mail: alexeitikhomirovprof@gmail.com

O. Berestneva
National Research Tomsk Polytechnic University, Tomsk, Russia
e-mail: ogb6@yandex.ru

A. Rossodivita
Luigi Sacco Academic Hospital, Milan, Italy
e-mail: italiancare@gmail.com

B. Gonçalves et al. (eds.), *Complex Networks VIII*,
Springer Proceedings in Complexity, DOI 10.1007/978-3-319-54241-6_15

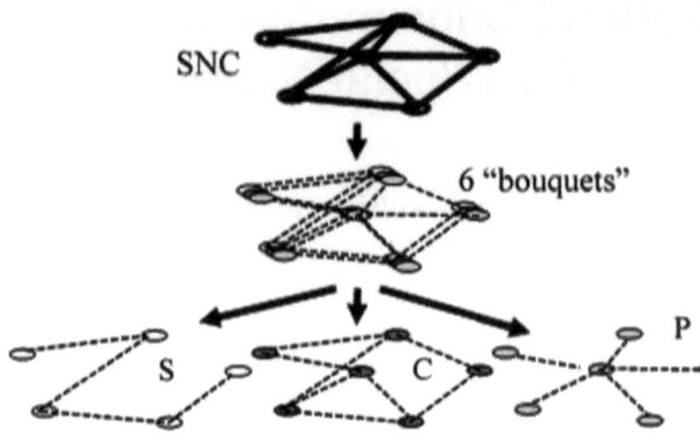

Fig. 1 Social network compositions (SNC) stratified into of *S*-, *C*-, and *P*-components

reveal details of human behavior from huge amount of tangled data [2–5]. Complex networks [6] have been successfully applied for studying not only separate networks but multilayer [7, 8] and interconnected entities of diverse multicomponent structures [9, 10]: all these are covered with general term multiplex networks. In the current study, the modern social network compositions (SNC) include social networks (*S* networks) per se, systems of information sharing (communication components, C-networks), and tool platforms providing the processes of sharing (P-networks). Generally, each component of these socio-cyber systems has non-trivial characteristics to be in focus of many explorers [11–13]. Following the concept of [14], we suppose that actors from each of *S*-, *C*-, and *P*-ensembles, integrated in triples—"bouquets", form so-called combined (often interdependent) stem networks (Fig. 1).

In general single actor multiple nodes belonged to different layers in multiplex networks might form a stem. However, for this study there is no necessity to consider more than one layer case for each of the S-, C-, and P-networks. Thus such actors as a computer device, information resource, and individual comprise a bouquet of social network composition. Traditionally a separate complex network is described in terms of graph theory. Then, in graph G_q of q component of social network composition

$$G_q = \left(V_q, E_q\right) \quad (1)$$

where V_q—a set of nodes (vertices), E_q—a set of links (edges), and a set of V_q—includes all participants of information sharing processes in the component

$$q = \{S, C, P\} \quad (2)$$

As a rule for any network structure, four main topological risk problems are claimed [15]: (a) disintegration under random or coordinated attacks on complex networks; (b) cascading failures; (c) congestion; (d) spreading processes of malicious activities.

Components of social network composition can be also a subject to all these four troubles which involve breaches of information security and distortion of such properties as confidentiality, integrity and availability. Gordon and Loeb [16] and

Huang with coauthors [17] in their formal models considered economic aspects of information security and revealed features and importance of optimal investment into information security. Helbing in [18] stated topological measures to mitigate system risks. Nevertheless, researchers have been faced so far by a number of the intricate problems in information security of social networking compositions.

2 Model of Social Networking Security

Topological Risk. Probabilistic nature of the processes that bring damages allows to define risk, R—the main security measure—as [19]: $R = P \times D$, where P—probability of the successful attack, D—damage caused by impact of an attack. In the proposed model, similar to [20], the attacks are revealed through the detailed description of triplets—threats, vulnerabilities and counter-measures for separate nodes. It should be noted that network risk, threats, vulnerabilities, counter-measures and losses generally have structure-dependent character. As a classic attack on network structure is focused on removal of nodes which is a result of coordinated threat, one has for topological risk $\mathcal{R}$:

$$\mathcal{R} = \mathcal{P}_{tN} \times \mathcal{P}_{vN} \times (1 - \mathcal{P}_{cN}) \times \mathcal{L}_N \tag{3}$$

here $\mathcal{P}_{tN}, \mathcal{P}_{vN}, 1 - \mathcal{P}_{cN}$ corresponding probabilities of structural threat, vulnerability and overcoming of counter-measures, $\mathcal{L}_N$—cost of topological damage, caused by attack on node set $N_a \in V$.

Metrics of Structural Behavior in Protected Network. Within the research network disintegrations—structural losses L, are estimated by calculations of a portion g of the nodes disconnected with giant cluster after successful attacks which were carried out against targets—nodes: $L = 1 - g$. In case of emerging threats, with topological features close to real, the model gives means to investigate network system risks in more complex environment, if compare to traditional one. We will note, the approach allows varying $\mathcal{P}_{ti} \times \mathcal{P}_{vi}$ through the strategy of a threat source and data errors in topology of the network. The values of $\mathcal{P}_{vi} = 1 - \mathcal{P}_{ci}$, give probabilities of overcoming counter-measures—as protection of a network node. So, if follow [16], the probability $\mathcal{P}_{vi}$ of a successful attack on a node i is connected with the investment in line with power expressions. Contrary, in the study it was suggested that $\mathcal{P}_{vi}$ decreases exponentially with increasing of protection "wall thickness" d. This thickness is defined by a traditional package of security measures and is connected with the security investment F_i. Thus $d_i \approx iF_i$, and in this case:

$$\mathcal{P}_{vi} = \exp(-d_i) = \exp(\mu \times F_i) \tag{4}$$

where μ - a coefficient which sets efficiency of financial means.

Security Level of a Protected Node. Security level of a separate element i of a network is defined by the value:

$$s\mathscr{L}_i = (1 - \mathscr{P}_{vi}) = 1 - \exp(-\mu \times F_i) \quad (5)$$

Security Level of a Protected Network. We underline one should distinguish security level of a separate node with that of a network in whole with security investment sum $\mathscr{F} = \sum_n F_i, n$—is power of a set V. It seems reasonable to determine axiomatic parameter security level of a network structure- by probability that any of elements won't be successfully attacked. If the probability to choose an attacked element i is $1/n$ then:

$$SL = 1 - \sum_n \mathscr{P}_{vi} = 1 - \left[\sum_n \exp(-\mu \times F_i)/n\right] \quad (6)$$

3 Findings

Analysed networks In this paper we study some networks of S-, C-, and P-character in the field of coordinated (i.e. intentional) threats to understand their information security investment sensitivity. For this research we concentrated our choice on the following real networks (see Table 1):

m is number of links, $\langle k\rangle$—average connectivity of nodes, γ connectivity degree in a power distribution of node connectivity.

The fact that the social networking components are organizationally inclined to protect "important" nodes and links were taken into consideration. For the modeling research two protective strategies are considered.

Protective Strategy 1. All nodes are protected (financially) equally: $F_i = F1_i = Const1 = 1/n$

Protective Strategy 2. In practice In practice nodes are protected unequally. Because the strategy of security determines distribution of financial resourses

Table 1 Social networks (S), communication networks (C), and networks of computer platform (P)

Type	Code	Name	n	m	$\langle k\rangle$	γ	Reference
P	CA	CAIDA	26,475	53,381	4.0326	~2	[21]
P	AS	Route views	6,474	13,895	4.2926	~2	[21]
C	HA	Haggle	274	28,244	206.16	1.5	[22]
S	FB	Facebook	63,731	817,035	25.640	~3	[23]
S	AP	Astro Physics collaboration	18,772	198,110	21.107	~3	[21]
S	JZ	Jazz musicians	198	2,742	27.697	5.3	[24]

between nodes, we investigated reaction of networks to threats in case when investment into protection of nodes is proportional to their connectivities. The total volume of investment into protection is the same, as for Strategy 1:

$$F2_i(k_i) = Const2 \times k_i/\mu \qquad \sum_{i=1}^{|V|} F1_i = \sum_{i=1}^{|V|} F2_i(k_i) = \mathscr{F} \tag{7}$$

It was supposed, as before, that the offensive party carries out the intentional successive choice of targets-nodes with highest connectivity. Certainly, a violator plans actions on disintegration with expectation of maximum effect at a minimum of expenses. Anticipating these threats, the defensive party builds protection, putting investment into its means so that not to exceed limited loss.

Comparison of Structural Loss in *S*-, *C*-, **and** *P*-**networks**. P-components of social network compositions-computer networks CA and AS both have properties of scale-free networks with power degree ~2, and low value of average connectivity ~4. We find that removal more than 1% of nodes causes essential damage to these unprotected network structures. It is possible to expect that *S*- and *C*-networks, which possess greater value of average connectivity, also have smaller structural vulnerability. Dependences of structural loss L in real *S*-, *C*-, and *P*-networks countering to destruction of 10% nodes are presented in Fig. 2. Values of the necessary financial volumes of protection measures which are uniformly distributed among nodes (in $1/\mu$ units) to provide necessary network security level SL are manifested as well. Results of calculations confirm that intentional threats of SNC disintegration are especially dangerous concerning a computer component (CA and AS networks). Loss as functions of security level for social networks FB, AP and JZ has similar behavior. Communication network HA demonstrates its intermediate character. Estimations of topological loss and protection costs for networks with 20% of nodes—targets are given on Fig. 3. Those have the distribution of protection investment proportional to node degree (Strategy 2). Comparison of results shown on the figures indicates that the strategy with protection of nodes in proportional dependence on connectivity

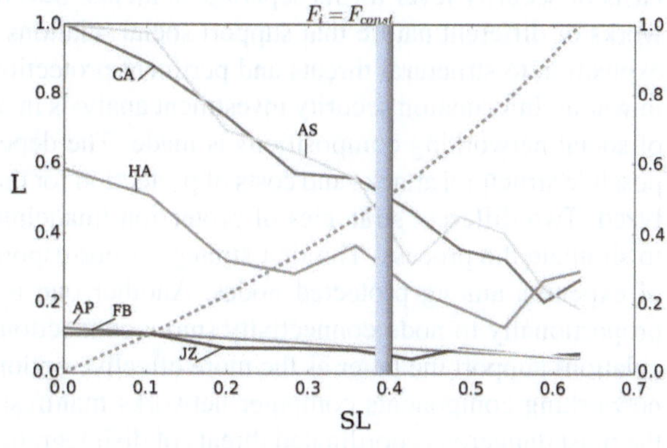

Fig. 2 Structural loss in network security (*L, solid line*) and security investment (*F, dashed line*) at different network security level (*SL*) (case of equally protected nodes)

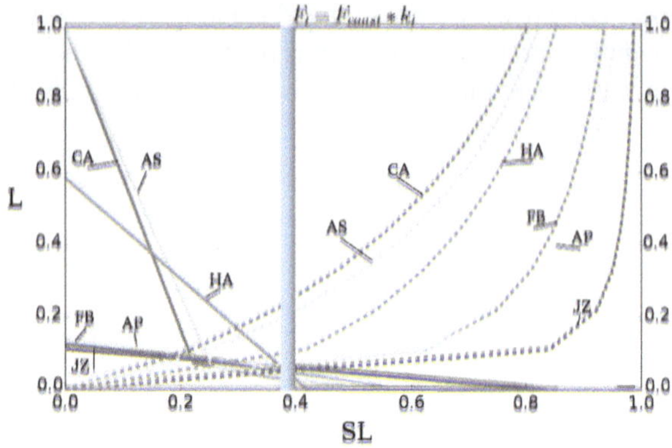

Fig. 3 Structural loss (*L*, *solid line*) and financial volumes (*F*, *dashed line*) for network protection as functions of network security level according to Strategy 2

(i.e. $F_i \sim k_i$) is more effective, than the strategy with uniform distribution of investment. And it is clear as counter-measures reduce probability of inactivation of the nodes representing the main targets for the classical strategy of coordinated threats. Also it should be underlined that one might observe a limit of network security level which is of sense to reach. The optimal value of the *SL* never exceeds 0.4 for both strategies of network protection (see Figs. 2 and 3).

4 Conclusions

The undertaken research is concluded to consider contemporary social networking constructions as multi-structural aggregates composing of: computers, communications and social networks. The model and the program tools are developed for estimations of topological risks for the networks which elements nodes—are provided with protection depending on the volumes of financing. A security description for a network is done using consideration of threats, vulnerabilities and countermeasures for individual nodes. The concept of network security level is designed on the basis of security level for its separate elements. Several representatives of real networks of different nature that support social relations are selected to simulate their exposition to structural threats and pertinent protections of nodes and the networks in whole. Information security investment analysis in case of real versatile examples of social networking compositions is made. The dependencies of losses caused by possible structural attacks and costs of protection for these selected networks are analyzed. Two different strategies of protection financing are taken into consideration to simulate the process. The first strategy is corresponded to a uniform distribution of expenses among protected nodes. Another one is implied dividing the budget proportionally to node connectivity (more connections more investment). The calculations support the latter as the more effective option. We show that among social networking components computer networks manifest their greatest sensitivities to the most dangerous coordinated threats of disintegration. In addition, it is found that network security level with optimal investment does not exceed $0.4 \sim 1/e$ for both strategies of network protection. It seems tempting to compare this limit value with the Gordon-Loeb rule that the optimal amount of investment to spend on information security is lower than $0.37 \sim 1/e$ of the expected loss following to a security breach [16]. As a result, the proposed model and its tools will allow to cover effectively topological problems of information security economics within the framework of the modern network information systems.

References

1. Boyd, D.M., Ellison, N.B.: Social network sites: definition, history, and scholarship. J. Comput.-Mediat. Commun. **13**, 210230 (2008)
2. Butts, C.T.: The complexity of social networks: theoretical and empirical findings. Social Netw. **23**(1), 3172 (2001)
3. Basim, M., Menezes, R.: The role of human relations and interactions in designing memory-related models for sensor networks. Sens. Transducers **199**(4), 42–51 (2016)
4. Grabowicz, P.A., Ramasco, J.J., Gonalves, B., Eguiluz, V.M.: Entangling mobility and interactions in social media. PLoS One **9**, E92196 (2014)
5. Szell, M., Sinatra, R., Petri, G., Thurner, S., Latora, V.: Understanding mobility in a social petri dish. Sci. Reports **2**, 457 (2012)
6. Barabsi, A.-L., Albert, R., Jeong, H.: Mean-field theory for scale-free random networks. Physica A **272**, 173187 (1999)
7. Kivela, M., Arenas, A., Barthelemy, M., et al.: Multilayer networks. J. Complex Netw. **2**(3), 203271 (2014)
8. Omodei, E., De Domenico, M., Arenas, A.: Evaluating the impact of interdisciplinary research: a multilayer network approach. Netw. Sci. **1**, 12 (2016)
9. Chen, Y.-Z., Huang, Z.-G., Zhang, H.-F., et al.: Extreme events in multilayer, interdependent complex networks and control. Sci. Reports **5**, 17277 (2015)
10. Wellman, B.: Computer networks as social networks. Science **293**, 2031–2034 (2001)
11. Landon, B.E., Keating, N.L., Barnett, M.L., et al.: Variation in patient-sharing networks of physicians across the United States. Jama **308**(3), 265–273 (2012)
12. Romero, D.M., Uzzi, B., Kleinberg, J.: Social Networks Under Stress WWW 2016, April 1115, 2016, Montral, Qubec, Canada. ACM 978-1-4503-4143-1/16/04 (2016)
13. Herrera, J.L., Srinivasan, R., Brownstein, J.S., et al.: Disease surveillance on complex social networks. PLoS Comput. Biol. **12**(7), e1004928 (2016)
14. Ashurova, Z., Myeong, S., Tikhomirov, A., et al.: Comprehensive mega network (CMN) platform: Korea MTS governance for CIS case study. In: Berestneva, O., Tikhomirov, A., Trufanov, A. (eds.) ITSMSSM 2016: Information Technologies in Science, Management, Social Sphere and Medicine Tomsk, May 2016, pp. 266–269. Atlantis Press (2016)
15. Regan, E.R.: Networks: Structure and Dynamics. Beth Israel Deaconess Medical Center, Department of Medicine, Boston (2002). http://regan.med.harvard.edu/Teaching/CVBR/Ravasz_Networks.pdf
16. Gordon, L., Loeb, M.: The economics of information security investment. ACM Trans. Inf. Syst. Secur. **5**(4), 438457 (2002)
17. Huang, D., Behara, R., Goo, J.: Optimal information security investment in a healthcare information exchange: an economic analysis. Decision Support Syst. **61**, 1–11 (2014)
18. Helbing, D.: Globally networked risks and how to respond. Nature **497**, 5159 (2013)
19. Cox Jr., L.A.: Some limitations of risk = threat? Vulnerability? Consequence for risk analysis of terrorist attacks. Risk Anal. **28**(6), 1749–1761 (2008)

20. Plum, M.M., Phillips, J., McCabe, P.H., et al.: Novel threat-risk index using probabilistic risk assessment and human reliability analysis. Report No. INEEL/EXT-03-01117, Idaho National Engineering and Environmental Laboratory, Idaho Falls, Idaho (2004). http://inldigitallibrary.inl.gov/sti/2535260.pdf
21. Leskovec, J., Kleinberg, J., Faloutsos, C.: Graph evolution: densification and shrinking diameters. ACM Trans. Knowl. Discov. Data **1**(1), 1–40 (2007)
22. Chaintreau, A., Hui, P., Crowcroft, J., et al.: Impact of human mobility on opportunistic forwarding algorithms. IEEE Trans. Mobile Comput. **6**(6), 606–620 (2007)
23. Viswanath, B., Mislove, A., Cha, N., Gummadi, K.P.: On the evolution of user interaction in Facebook. In: WOSN09: Proceedings of Workshop on Online Social Networks, August 17, 2009, Barcelona, Spain (2009). http://www.mpi-sws.org/gummadi/papers/wosn23-viswanath.pdf
24. Gleiser, P.M., Danon, L.: Community structure in jazz. Adv. Complex Syst. **6**(4), 565–573 (2003)

Part V
Social Structure

Emergence of Social Balance in Signed Networks

Andreia Sofia Teixeira, Francisco C. Santos and Alexandre P. Francisco

Abstract Social media often reveals a complex interplay between positive and negative ties. Yet, the origin of such complex patterns of interaction remains largely elusive. In this paper we study how third parties may sway our perception of others. Our model relies on the analysis of all triadic relations taking into account the influence and relations with common friends, through large-scale simulations. We show that a simple peer-influence mechanism, based on balance theory of social sciences, is able to promptly increase the degree of balance of a signed network—with balance defined as the fraction of positive cycles—irrespectively of the network we start from. Additionally, our results indicate that the tendency towards a balanced state also depends on the network connectivity and on the initial distribution of signs.

Keywords Balance theory · Network analysis · Social networks

1 Introduction

Signed networks are networks where the links have a sign expressing some positive or negative tie between individuals [1–8]. It is well-known that in social networks one can be friendly or unfriendly with others and that this can change over time. Moreover, individuals also shape and reshape their social environment themselves and are responsible for the specific features that characterize their social network [9–12]. Social balance theory, a concept developed by Heider [13], and later adapted to a graph-theoretic model by Cartwright and Harary [1], states that in a triad, the relations of friend-enemy tend to converge to two balanced states: "the friend of my

A.S. Teixeira (✉) · F.C. Santos · A.P. Francisco
INESC-ID/Instituto Superior Técnico, Universidade de Lisboa, Lisbon, Portugal
e-mail: sofia.teixeira@tecnico.ulisboa.pt

F.C. Santos
e-mail: franciscocsantos@tecnico.ulisboa.pt

A.P. Francisco
e-mail: aplf@tecnico.ulisboa.pt

B. Gonçalves et al. (eds.), *Complex Networks VIII*,
Springer Proceedings in Complexity, DOI 10.1007/978-3-319-54241-6_16

friend is my friend" and "the enemy of my enemy is my friend", otherwise there will be tension between them.

In 1946, Fritz Heider published an initial study about how affective ties—as to like, to love, to esteem, etc., and their opposites—would influence interpersonal relations [13]. These simple cognitive configurations between people and objects led to the conclusion that a triad is balanced if the three links are positive, or if two are negative and one positive, otherwise tension would emerge. This was a primary approach to social balance. Later, Cartwright and Harary, extended this notion of balance to a graph—structural social balance—and used the concept of signed graphs, where the ties between the individuals have a positive or a negative sign, to express those kind of relations [1, 2]. They extended the concept of triad to a cycle, allowing cycles with more than three edges, and defining the sign of the cycle as the product of the signs of its edges. A cycle is then considered balanced if the product is positive. They also introduced the concept of degree of balance of a signed network as the ratio of the number of positive cycles to the total number of cycles. Let G be a signed graph, $c(G)$ be the number of cycles of G, $c_+(G)$ be the number of positive cycles of G, and $b(G)$ be the degree of balance of G. Then:

$$b(G) = \frac{c_+(G)}{c(G)}$$

In our work we use this measure applied to triads, that is, cycles of size 3.

Following the work of Cartwright and Harary, in 1967, Davis [3] studied the relation between clustering and structural balance in graphs. The main question was about what conditions were necessary and sufficient for the graph to be separated into two or more subsets of nodes, where each positive edge would link two nodes of the same subset and a negative edge would link nodes from different subsets. Those conditions were: a signed network is clusterable if and only if the network does not contain any cycle with exactly one negative link. This introduced the notion of *weak balance theory* as it allows for cycles/triads to have all signs negative, meaning that "the enemy of my enemy can be an enemy", allowing more than two subsets to be created. The main conclusion was that all balanced graphs are clusterable.

Global structural balance has also been studied. Doreian et al. [4] created an agent-based simulation model based on two levels: a micro-level that explores Heider's theory at an individual level, to minimize individual tension; a macro-level that explores Cartwright and Harary's at a group level dynamics. This simulation model is only for small groups dynamics as the designed variables have complicated impacts. Facchetti et al. [7] implemented an algorithm for ground-state calculation in large-scale Ising spin glasses, to compute the global level of balance in large undirected networks. And recently Estrada and Benzi [8] published a study about structural social balance in directed networks.

In this work, we evaluate how the relations between individuals change over time, based on the relations with common friends, and if those changes converge to a balanced social structure. We present a simulation model that, at each iteration, evaluates if the sign between two individuals must change to minimize tension across

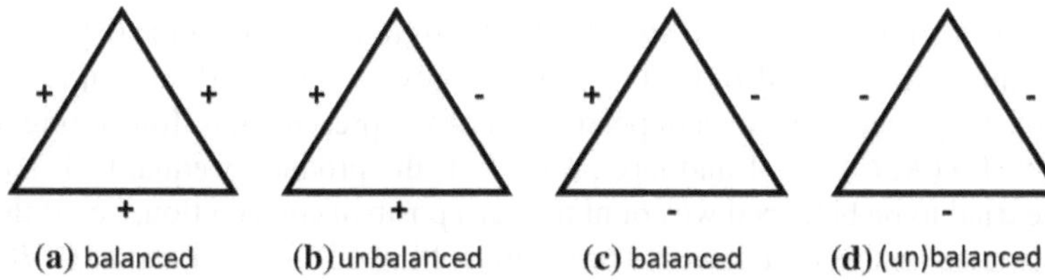

Fig. 1 Social Balance Theory, by Cartwright and Harary [1]. The triads are considered balanced if the product of the signs are positive. Davis introduced the weak balance structure that considers all triads but the second to be balanced

triads. A triad is considered balanced if its edges have the signs {+, +, +} or {+, −, −}, meaning that the product of its signs has to be positive—see Fig. 1. We consider that the polarity of the relations is reciprocal, considering only undirected networks. We run our simulations with the original distribution of signs of each chosen dataset, but also with a random distribution of the signs, both in the same proportion as in the original network and in equality proportion of positive and negative links. We observe that the evolution of the signs between individuals involved in triads converge towards an increase of structural balance, minimizing the tension between the individuals, but also that the final dominant triads depend on the initial distribution of the signs.

2 Methods

Let $G = (V, E)$ be an undirected and weighted graph (signed social network), with $n = |V|$ vertices (individuals) and $m = |E|$ edges (ties), and with edges weight between two individuals (a,b): $w(a, b) = w(b, a) = 1$, if it is a positive tie, $w(a, b) = w(b, a) = -1$ if it is a negative tie. For each pair of individuals with friends in common, our model will count how many of those relations contribute with a positive or negative sign, based on balance theory.

Looking into the example illustrated in Fig. 2: given a network let us consider individuals A and B that have C, D, E, F, G, H as friends in common. We now evaluate if the product between $w(A, C)$ and $w(C, B)$ is positive or negative, and the same for the other neighbours. Because we want to reduce tension in triads, the sign between individuals A and B will depend on a majority count between positive and

Fig. 2 What will be the sign between A and B? It will depend on the majority of the signs of the products of each vertex A and B with each neighbour

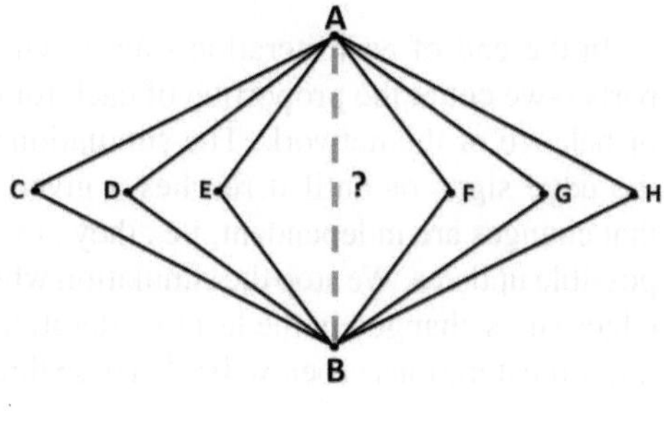

negative products of the other relations in the triads related to that link. Given the sign between individuals A and B, $w(A, B)$, it will only be updated if the majority of the counts of the products have an opposite sign of the present sign. Illustrating a little bit more: if $w(A, C) = -1$ and $w(C, B) = -1$, the product is equal to 1, so if we want the triad to be balanced we count this as a positive contribution, i.e., if the sign only depended on this triad it would be positive. If $w(A, C) = -1$ and $w(C, B) = 1$, the product is equal to -1, so if we want the triad to be balanced $w(A, B)$ would have a negative contribution in the count. If the sign only depended on this triad $w(A, B)$ would be negative. We remind that a triad is balanced if the product of the signs of its edges is positive. We do this count for each neighbour in common. In other words: $w(A, B)$ will be the sign corresponding to the majority of positive or negative contribution counts.

The algorithm runs in two parts, as follows:

```
for each user u do
    for each neighbour n do
        collect the friends in common

        for each friend in common c do
            if the product between w(u, c) and w(n, c) == 1 then
                pos(u, n) ⟵ pos(u, n) + 1
            else
                neg(u, n) ⟵ neg(u, n) + 1
            end if
        end for
    end for
end for
for each edge (a,b) do
    if pos(a,b) == neg(a,b) then
        there is no update and w(a, b) stays the same
    end if
    if pos(a,b) > neg(a,b) then
        w(a, b) ⟵ 1
    else
        w(a, b) ⟵ −1
    end if
end for
```

In the end of each iteration—an iteration corresponds to the execution of both parts—we count the proportion of each four possible triads and calculate the degree of balance of the network. The simulations run until there are no more changes in the edge signs or until it reaches a given threshold on the counts changes. Note that changes are independent, i.e., they are synchronous and do not depend on other possible updates. We stop the simulation when either the average of the fraction of the edges signs changed in the last two iterations is below 10^{-2}, or its difference for the last three iterations is below 10^{-4}. These thresholds were determined experimentally.

Table 1 Networks used in the simulations

Network	# Nodes	# Edges	% Edges +	% Edges −	# Triangles
HighlandTribes	16	58	50.00	50.00	68
Epinions	131 828	708 507	83.25	16.74	4770102
Slashdot	82 144	498 532	76.41	23.59	571127

3 Results and Discussion

In these experiments we used well-known signed social networks: Highland Tribes, the signed social network of tribes of the GahukuGama alliance structure of the Eastern Central Highlands of New Guinea, from Kenneth Read (1954). The network contains sixteen tribes connected by friendship and enmity[1]; Epinions, a who-trust-whom online social network of a general consumer review site Epinions.com[2]; and Slashdot, a website which allows users to tag each other as friends or foes.[3] We also created cliques with different sizes just to compare complete connected networks with Epinions and Slashdot that are large-scale sparse networks. Because Epinions and Slashdot datasets are directed networks, we performed some operations in these networks to make them undirected. We analysed each network and if some relation had a conflict—one edge in one direction positive, and in the other direction negative—we removed that edge, keeping only the relations that are reciprocal.

In Table 1 we can find the characteristics of each network. We processed each social network in three different ways: (1), we started by running the simulations with the networks as they were after removing conflicting edges; (2), for each network we randomly distributed the signs of the edges in the same proportion as in the original network; (3) for each network we distributed randomly and evenly positive and negative signs, i.e., 50% of positive edges and 50% of negative.

In Fig. 3 we present the results of the simulations. It contains the initial and final distribution of the four possible triads and of the degree balance. As we can observe, the initial distribution of triads in the *Random* and in the *Evenly* networks are very different when compared to the original. Even when maintaining the initial proportion of positive and negative links, this means that there are some triads that are overrepresented in the original network, which indicates that the way signs are distributed initially has direct impact in the structural balance.

We can observe that having a dominant quantity of positive or negative links makes a network to converge to a dominant all-positive triads: (1) if individuals like each other and the friends in common also like each other, then there is no social reason to change; (2) if individuals do not like each other or the friends in common, then there are triads in tension and the update rule forces signs of all-negative triangles

[1] http://konect.uni-koblenz.de/networks/ucidata-gama.

[2] https://snap.stanford.edu/data/soc-sign-epinions.html.

[3] https://snap.stanford.edu/data/soc-sign-Slashdot090221.html.

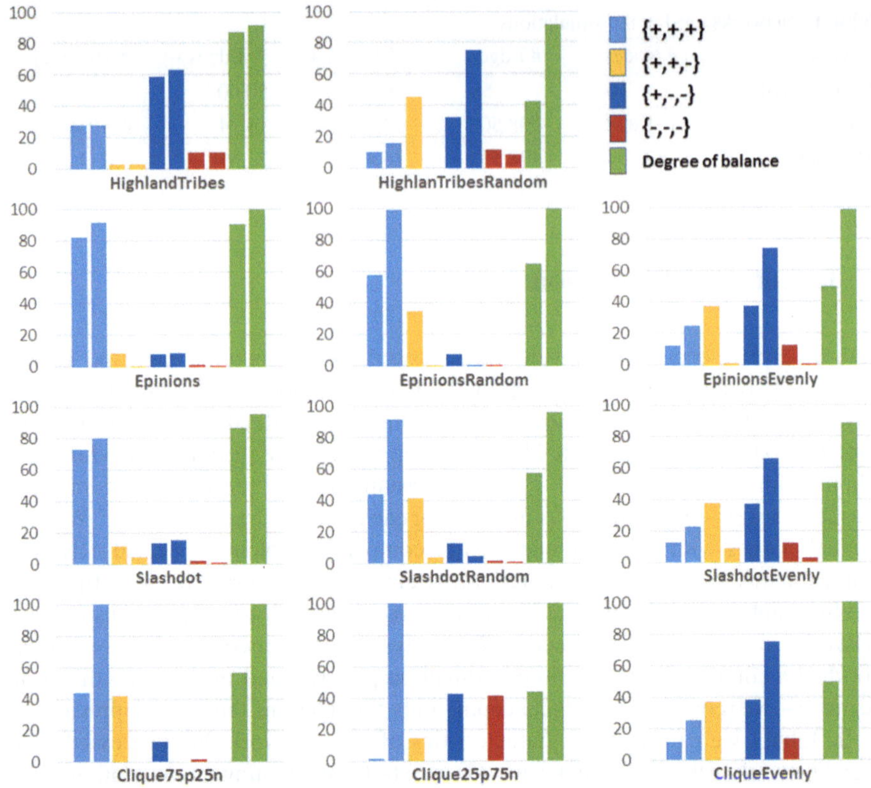

Fig. 3 Simulations results. Each pair of columns corresponds to the initial and final distribution of each triad, except for the last pair which represents the initial and final degree of balanced of the network. As seen in Fig. 1, only the first and third triads are considered balanced. *Random* means that the signs were distributed randomly in the same proportion of the original network and *Evenly* means that the signs were distributed randomly with 50%–50% of positive-negative signs. HighlandTribes does not have Evenly because the distribution is already 50%–50%. We omitted the size of the clique, but we used sizes between 8 and 64 and the results were the same

to change to positive. In all networks there is an increase of balanced triads, but depending on the initial distribution of signs, the dominant triads are different. All networks with 50%–50% of positive and negative signs, instead of converging to the same initial dominant triads, converge to the two-negative one-positive triads. There are two reasons for this to happen: we have a high initial distribution of the triad $\{+, +, -\}$ that will always change to $\{+, -, -\}$; the decrease of the distribution of the triad $\{+, +, +\}$, when comparing to the original networks, also indicates that there is not enough all-positive triads to compete with the new dominant $\{+, -, -\}$. This leads to the conclusion that initial distribution of positive and negative ties has direct impact in how signs can evolve and, again, in the degree of balance.

Making the signs evolve based on the balance theory criteria will always force the individuals of the network to act towards a minimization of tension. There is a strong tendency towards balance, but not always enough to achieve 100% balance. This happens in the non-fully connected networks, usually when the changes reach the threshold on the number of changes. We observe that the achievement of total degree of balance may depend on the connectivity of the network—fully connected networks eventually converge as can be seen in cliques. This last conclusion was already derived theoreticly in previous works by Antal [14] and Arnout van de Rijt [15]. With different approaches, both come to conclusion that in a complete connected networks, when updating triads with the goal of minimizing imbalance, a balanced state is achieved.

4 Conclusions

Network Science [13, 16] has provided key insights on how individual states, from individuals choices [17–19], epidemic states [20], strategic behaviours [9, 21–24], and opinions [25], among other traits, are locally influenced by their social ties and by the overall topology of interactions within a population. While the dynamics at the level of nodes is crucial, analogous dynamics occurs at the level of states and weights of links [5, 6, 26, 27], with particular relevance within social settings.

In this context, the study of signed networks has benefited enormously from the quick growth of data on online social networks and more models are needed to understand its particular dynamics. In this work we report the results of a simulation approach to understand how the signs of the networks can evolve taking into account the social theories of structural balance and dynamics of peer-influence. We observed that updating a relation between two individuals based on the relations between both and the friends in common (triads) have impact in the evolution of the structural balance, but we also noticed that the way the signs of the network are initially attributed to each relation is determinant. The principles outlined in the proposed update rule for signs of links are general enough to be applied to other dynamical processes occurring in static and dynamic networks, where the sign (or weights) of ties plays an important role, from spreading of contagious diseases to diffusion of information in social networks.

For future work we plan to extend this study applying other sign distributions and update rules, including a probability approach similar to Antal [14] approach, but with the probabilities being proportional to the positive/negative counts explained in this work.

Acknowledgements This work was partly supported by national funds through Universidade de Lisboa and FCT—Fundação para a Ciência e Tecnologia, under projects TUBITAK/0004/2014, PTDC/EEI-SII/5081/2014, PTDC/EEI-SII/1973/2014, PTDC/MAT/STA/335 8/2014 and UID/CEC/50021/2013.

References

1. Harary, F.: On the notion of balance of a signed graph. Mich. Math. J. **2**, 143–146 (1954)
2. Cartwright, D., Harary, F.: Structural Balance: a Generalization of Heider's theory. Psychol. Rev. **63**, 277–292 (1956)
3. Davis, J.A.: Clustering and structural balance in graphs. Hum. Relat. **20**, 181–187 (1967)
4. Hummon, N.P., Doreian, P.: Some dynamics of social balance processes: bringing Heider back into balance theory. Soc. Netw. **25**(1), 17–49 (2003)
5. Leskovec, J., Huttenlocher, D., Kleinberg, J.: Signed networks in social media. In: 28th ACM Conference on Human Factors in Computing Systems (CHI) (2010)
6. Leskovec, J., Huttenlocher, D., Kleinberg, J.: Predicting positive and negative links in online social networks. In: 28th ACM Conference on Human Factors in Computing Systems (CHI) (2010)
7. Facchetti, G., Iacono, G., Altafini, C.: Computing global structural balance in large-scale signed social networks. Proc. Natl. Acad. Sci. USA **108**, 20953–20958 (2011)
8. Estrada, E., Benzi, M.: Are social networks really balanced? arXiv:1406.2132 [physics.soc-ph] (2014)
9. Skyrms, B.: The Stag Hunt and the Evolution of Social Structure. Cambridge University Press (2004)
10. Kossinets, G., Watts, D.J.: Empirical analysis of an evolving social network. Science **311**(5757), 88–90 (2006)
11. Santos, F.C., Pacheco, J.M., Lenaerts, T.: Cooperation prevails when individuals adjust their social ties. PLoS Comput. Biol. **2**(10), e140 (2006)
12. Gross, T., Blasius, B.: Adaptive coevolutionary networks: a review. J. R. Soc. Interface **5**(20), 259–271 (2008)
13. Barabási, A.-L.: Network Science. Cambridge University Press (2016)
14. Antal, T., Krapivsky, P.L., Redner, S.: Dynamics of social balance on networks. Phys. Rev. Lett. **72**, 036121 (2005)
15. van de Rijt, A.: The micro-macro link for the theory of structural balance. J. Math. Sociol. **35**, 94–113 (2011)
16. Dorogovtsev, S.N.: Lectures on Complex Networks. Oxford University Press, Oxford (2010)
17. Christakis, N.A., Fowler, J.H.: The spread of obesity in a large social network over 32 years. N. Engl. J. Med. **357**(4), 370–379 (2007)
18. Christakis, N.A., Fowler, J.H.: The collective dynamics of smoking in a large social network. N. Engl. J. Med. **358**(21), 2249–2258 (2008)
19. Pinheiro, F.L., Santos, M.D., Santos, F.C., Pacheco, J.M.: Origin of peer influence in social networks. Phys. Rev. Lett. **112**(9), 098702 (2014)
20. Pastor-Satorras, R., Castellano, C., Van Mieghem, P., Vespignani, A.: Epidemic processes in complex networks. Rev. Mod. Phys. **87**(3), 925 (2015)
21. Santos, F.C., Pacheco, J.M., Lenaerts, T.: Evolutionary dynamics of social dilemmas in structured heterogeneous populations. Proc. Natl. Acad. Sci. USA **103**(9), 3490–3494 (2006)
22. Santos, F.C., Santos, M.D., Pacheco, J.M.: Social diversity promotes the emergence of cooperation in public goods games. Nature **454**(7201), 213–216 (2008)
23. Szabó, G., Fath, G.: Evolutionary games on graphs. Phys. Rep. **446**(4), 97–216 (2007)
24. Rand, D.G., Nowak, M.A., Fowler, J.H., Christakis, N.A.: Static network structure can stabilize human cooperation. Proc. Natl. Acad. Sci. USA **111**(48), 17093–17098 (2014)
25. Castellano, C., Fortunato, S., Loreto, V.: Statistical physics of social dynamics. Rev. Mod. Phys. **81**(2), 591 (2009)
26. Barabasi, A.-L.: The origin of bursts and heavy tails in human dynamics. Nature **435**(7039), 207–211 (2005)
27. Onnela, J.-P., et al.: Structure and tie strengths in mobile communication networks. Proc. Natl. Acad. Sci. USA **104**(18), 7332–7336 (2007)

Community Detection in the Network of German Princes in 1225: A Case Study

S.R. Dahmen, A.L.C. Bazzan and R. Gramsch

Abstract In the context of historical research, clustering of different groups into warring factions can lead to a better understanding of how conflicts arise or can be avoided. Using a spin-glass-based community detection algorithm, we study the crisis of 1225 between the Emperor of the Holy Roman Empire Frederick II and his son Henry VII, which almost led to a dissolution of the empire. Our main goal is to see how good this method is in detecting this rift when compared to the results of an analysis performed by one of the authors (Gramsch) using standard social balance theory applied to history.

1 Introduction

One of the main tasks in network theory is the detection of communities. The question whether or not a network can be partitioned into clusters is not trivial and it is contingent on the question being asked. There are many criteria on how a community can be defined and detected (see [4] for an extensive review on the subject). In the context of social networks in general and historical networks in particular, clustering can have far-reaching consequences, especially when clusters are involved in conflicts. Under a sociological perspective, a natural way of grouping nodes is that of social balance theory, a model of human relationships that can be traced back to the works of F. Heider on cognitive dissonance theory [8]. It is built upon the notion that, in a triad of nodes, the positive or negative relation between two nodes is reflected in

S.R. Dahmen (✉)
Instituto de Física da UFRGS, Porto Alegre 91501–970, Brazil
e-mail: silvio.dahmen@ufrgs.br

A.L.C. Bazzan
Instituto de Informática da UFRGS, Porto Alegre 91501–970, Brazil
e-mail: bazzan@inf.ufrgs.br

R. Gramsch
Historisches Institut der Universität Jena, Fürstengraben 3, 07743 Jena, Germany

B. Gonçalves et al. (eds.), *Complex Networks VIII*,
Springer Proceedings in Complexity, DOI 10.1007/978-3-319-54241-6_17

their relation to the third node (see Sect. 3.1). In order to test this idea in a historical setting, one of the authors studied the conflict that arose between the years of 1225 and 1235 in the Holy Roman Empire, a conflict which pitted the Emperor Frederick II against his heir, Henry VII [5]. Based on Heider's theory, Gramsch showed that the dispute led to a rift among the prince-electors, thus threatening the stability of the empire [5].

The main goal of this paper is to use a clustering algorithm for this event, considering the role of negative links, and compare it to the results found by Gramsch. Far from trying to rewrite history anew, since historical events are extremely complex, spanning years and sometimes thousands of players, our goal is rather humble: to see if network analysis, particularly community detection, may be used as a viable tool to help historians see patterns which otherwise could not be seen.

This paper is organized as follows: we first give a brief overview of the crisis of 1225–1235 within the Holy Roman Empire. In Sect. 3, we present materials and methods. We then discuss the results obtained by a traditional historical analysis and show how a spin-glass-based community detection algorithm compares with this analysis.[1]

2 Background and Related Work

In the present work we deal with particular aspects of the coalition and conflicting forces that underlie the reign of Henry VII in the Holy Roman Empire [5, 6]. In medieval times monarchic power was strongly restricted, and within the confines of the Holy Roman Empire, a coalition of many sovereigns, a consensus among rulers was extremely important for a successful rule of the elected Emperor. This became evident during the era of emperor Frederick II (1212–1250) and his son, King Henry VII (1220–1235). In 1235, due to the political incapacity of Henry, who sacked some princes of their power, Frederick II had to disavow his son, lest he cause further damage to the authority of the Staufian dynasty and lead to its demise. The conflict involved 68 sovereigns. Notwithstanding the complexity of relationships, Gramsch convincingly demonstrates that network analysis may provide new vistas on the overall structure of the conflict which lead to the deposition of Henry [5].

He depicted the political system of the medieval German empire as a network of princes, kings, counts, bishops and other sovereigns (henceforth called actors). Based on Heider's structural balance theory [8] (see Sect. 3.1), he was able to characterize not only the existence of a relationship between actors A and B but also that such relationships could be neutral, negative (hostile), or positive (friendly). The conflicts

[1] An extended version of this article with details on the spin-glass model can be found in http://xxx.lanl.gov and https://www.academia.edu/30801915/Community_Detection_in_the_Network_of_German_Princes_in_1225_a_Case_Study.

are of various natures. The existence of negative relations is essential for the method to work. They normally represent conflicts of various natures such as territorial or status competition, legal or military conflicts. Positive relationships in this context can be kinship or political alliances. The analysis was carried out over a period of ten years of political relations and interactions among actors (from 1225 to 1235). These form the so-called socio-matrices, which can be identified with adjacency matrices, albeit with negative entries. Gramsch's proposition is that within a cluster there should be no conflict among actors.

The most important feature is the dual structure in the network, where each group is separated by various conflicts. We recall that, previously, these conflicts were considered in isolation. However [6] showed that there were hidden relations between them. For instance, in 1225, emperor Frederick II predominantly collaborated with actors one group while Henry VII with opposing group. This then shows the origins of the later conflict between the father and the son.

Further, this analysis was able to show what happened between the years 1232 and 1235 (see figures in [6]), namely, which actors stayed together in one cluster, which ones changed political coalitions, and how the front line of conflicts changed geographically. In short, one can observe that the political situation in 1232 was characterized by an antagonism of two factions, each of which composed of two clusters. These two factions were, each, supported by Frederick and Henry, i.e., they favored different groups of princes. Between 1232 and 1234, Frederick decided to depose his son in order to avoid further consequences and recover the complete control over his empire. These two antagonistic factions then start to decay in 1233 and disappear almost completely by 1235.

3 Materials and Methods

In this section we discuss the main methods used in our approach: Heider's structural balance theory and the Potts Model. Following, we discuss their use for analyzing the network of 68 actors who take part in the historical event mentioned in Sect. 2.

3.1 Heider's Structural Balance Theory

In his seminal work of 1946 Heider asked the question about how an individual A's attitude towards B influences the way a third individual C relates to B. It originated the so-called structural balance theory, which states that a society is balanced when 'a friend's friend (enemy) is also my friend (enemy)'. If all triads of a network of relationships are balanced, the network is said to be balanced. The question naturally arises whether a network of individuals with such relationships can be grouped into separate communities or not. Harary [7] showed that if a connected network is balanced, it can be split into two opposing clusters. This was later generalized to cycles with more than 3 individuals, to the idea of a k-cycle [1, 3]. A network is k-balanced if it can be divided into k clusters where within each cluster there are only positive relationships. In real life, however, not all clusters are balanced. There will always be within a cluster of positive relations some nodes with negative ones. The number of such misplaced links is called 'frustration', a term borrowed from the physics of spin systems. The task is to find a configuration which minimizes frustration. The similarity between Heider's theory and a system of interaction spins led Reichardt and Bornholdt to introduce a method of community detection based on a mapping between a graph and a q-state Potts Model [9]. Their method was generalized by Traag and Bruggeman to account for the possibility of hostile links [10]. We describe their method below and would like to remark that other clustering algorithms cannot be used due to the presence of negative links.

3.2 Spin-Glass-Potts Model

The Potts model is a model of interacting spins where each spin can have q different values. The model is called spin-glass because spins are not spatially ordered (as in a crystal). Spins tend to align (or repel) themselves if they have the same (different) q. The attraction/repulsion is mediated by the Hamiltonian of the system, i.e. its energy for a given configuration $\{\sigma\} = \{\sigma_1, \sigma_2, \sigma_3, \ldots\}$ of clusters σ_1, σ_2 etc. Minimizing the Hamiltonian is equivalent to minimizing Frustration [10]. Given the adjancency matrix with elements A_{ij}, the Hamiltonian reads as in Eq. 1, where δ is Kronecker's delta function. The $p_{ij}^{\pm}$'s are the probabilities that links i and j are positively or negatively connected and γ are free parameters to tune the relative weight of positive and negative links.

$$H(\sigma) = -\sum_{i,j} \left[A_{ij} - (\gamma^+ p_{ij}^+ - \gamma^- p_{ij}^-) \right] \delta(\sigma_i, \sigma_j) \quad (1)$$

We refer the interested reader to [10] for more details on how to choose these probabilities.

3.3 Detecting Communities Using Spin-Glass

In order to detect the community structure for the conflict between Frederick and his son, we used the igraph implementation of the spin-glass algorithm (Python variant) [2]. Each actor is represented by an abbreviated name. As in [5], we use one socio-matrix (adjacency matrix) for each year (unless otherwise stated).

We use a set of adjacency matrices (prepared by R. Gramsch), where A_{ij} indicates whether or not there is a relationship between actors i and j, and, if there is, whether it is neutral, friendly, or hostile. Based on a suggestion of R. Gramsch we excluded all relationships involving Frederick II, Henry VII and the Pope, as these are the main actors of the conflict and served most of the time as liaisons between opposite groups. They introduce a bias in clustering, thus hiding important patterns. Results reported in the next section, thus, do not include these three actors. We remark that the same procedure was performed by Gramsch in his investigations; thus the results are comparable.

The spin-glass method needs as input the number n of communities. We chose $n = 2$, to see whether the method would lead to a partitioning of the network comparable to that found by Gramsch. If one gives a higher value of n, the method will produce n communities but normally for n above a certain threshold (in some of our cases 5 or above), the routine will give always at most 5 clusters, usually less.

4 Results

We have run the spin-glass with, as mentioned, the number of spins set to 2, producing thus partitions that should separate the conflicting parties. We did this for each year. Figures 1 and 2 show, for the sake of illustration, the clusterings for years 1225 and 1235 respectively.[2] Please notice the reduction of red edges (hostility) in the year 1235. Besides these edges, we have also yellow edges (neutral relationship) and black ones (friendly relationships).

In order to compare the quality of the clustering produced originally by Gramsch in [5, 6] with those from the spin-glass method, we use the Rand index, define in Eq. 2. In this equation a is the number of pairs of nodes that are in the same set in both partitions X, Y while b is the number of pairs that are in different sets in partition X and continue to be so in X; n is the number of nodes. A Rand index of 1 implies total agreement (clusters are identical) while a 0 implies total disagreement.

$$R = \frac{a + b}{\binom{n}{2}} \tag{2}$$

Table 1 shows the Rand indexes when we do a comparison, year by year, with the original partitioning of Gramsch. We remark that, since the spin-glass method is not deterministic, we ran spin-glass community detection 30 times for each year. Thus the table also shows the standard deviation associated with the mean value.

[2]We remark that, obviously, this is the result of a single run, thus different runs can produce slightly different partitions.

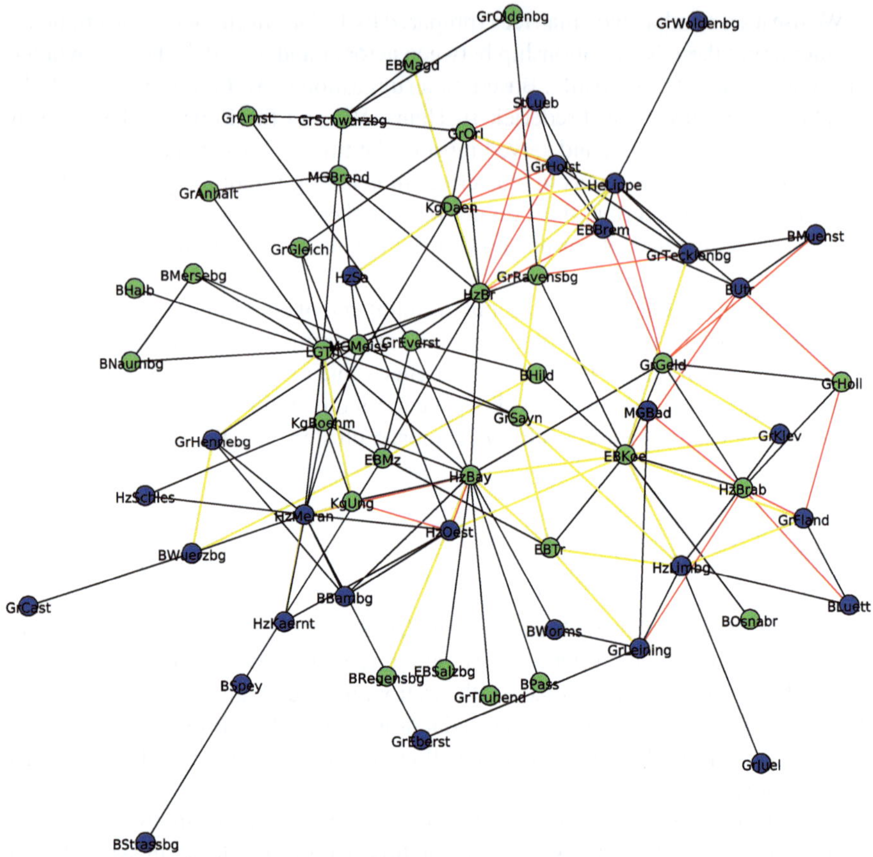

Fig. 1 The structure of the communities (clustering)—year 1225

The values, as can be seen in the table, indicate a good agreement between the spin-glassmethod and Gramsch's original partitioning, based on Heider's structural balance. We would like to point out that, for the year 1230, the agreement is comparatively low. This is due to the fact that in 1230 there occurred a temporary agreement between sovereigns. Quoting Gramsch ([5], p. 222): 'During the first quarter of the 1230, when peace talks between the Emperor and the Pope began, the sovereigns placed themselves in such a close [league] as it was never to be seen again: 58 joined into one coalition.' So for this year there is only one cluster. Since the method requires a a priori number of cluster to be created, which was set to 2, the Rand index is smaller and is about 0.5, which corresponds to the probability of placing nodes with a 50–50 change on each cluster.

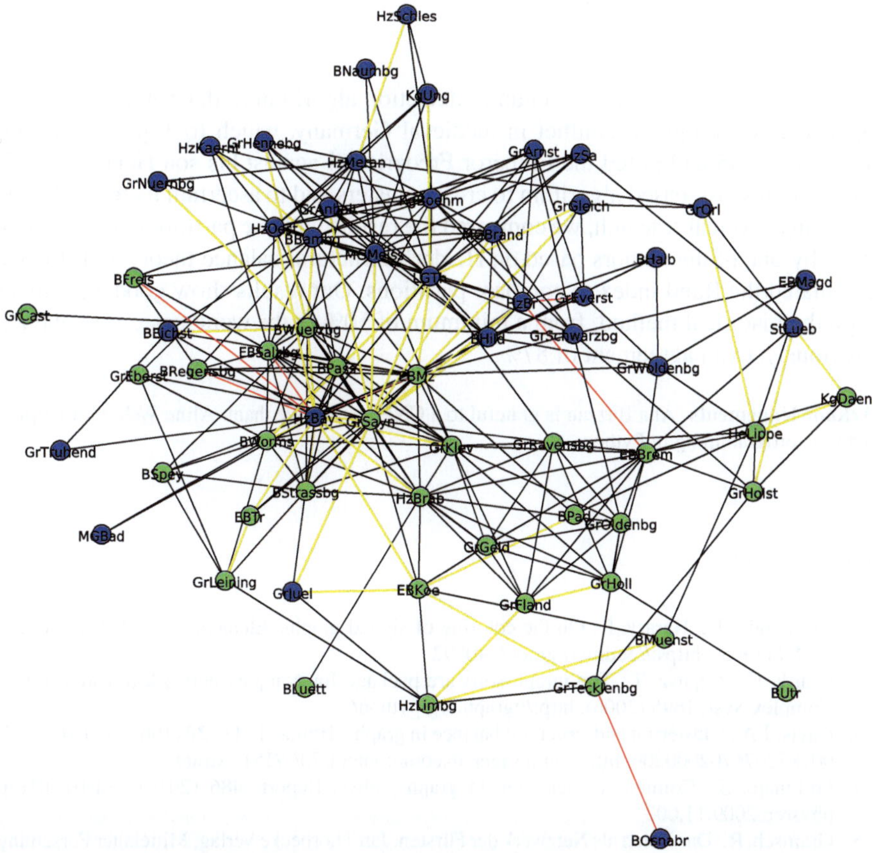

Fig. 2 The structure of the communities (clustering)—year 1235

Table 1 Rand indexes (mean and standard deviation), by year

Year	Rand index		Year	Rand index	
	Mean	St. dev.		Mean	Std. dev.
1225	0.78	0.06	1226	0.8	0.06
1227	0.66	0.15	1228	0.65	0.13
1229	0.73	0.05	1230	0.53	0.03
1231	0.84	0.9	1232	0.85	0.08
1233	0.87	0.09	1234	0.78	0.04
1235	0.87	0.04			

5 Conclusion

In this paper we applied a community detection algorithm to determine clusters of opposing sovereigns in conflict in medieval Germany, which took place between 1225 and 1235 and pitted the Emperor Frederick II against his son Henry VII. We used a spin-glass-based algorithm to create clusters and to ascertain its feasibility as a tool in historical research, we compared the results with the partitioning previously done by one of the authors based on Heider's structural balance theory. For this we calculated the Rand index to compare partitions. Our results show good agreement with the historical method, from a minimum of 50% in the worst case, as explained previously, to an agreement of 87%.

Acknowledgements Ana Bazzan is grateful to a CNPq grant. We thank Aline Weber for helping with the programming in Python.

References

1. Cartwright, D., Harary, F.: On the coloring of signed graphs. Elemente der Mathematik **23**, 85–89 (1968). https://eudml.org/doc/140892
2. Csardi, G., Nepusz, T.: The igraph software package for complex network research. Int. J. Complex Syst. 1695 (2006). http://igraph.org/python/
3. Davis, J.A.: Clustering and structural balance in graphs. Human Relat. **20** (1967). doi:10.1177/001872676702000206. http://hum.sagepub.com/content/20/2/181.extract
4. Fortunato, S.: Community detection in graphs. Phys. Reports **486** (2010). doi:10.1016/j.physrep.2009.11.002
5. Gramsch, R.: Das Reich als Netzwerk der Fürsten. Jan Thorbecke Verlag, Mittelalter Forschung (2013)
6. Gramsch, R.: Conflicts as a structure-forming force: the reign of Henry (VII) (1225–1235) in network-analytic perspective. In: Multiplying Middle Ages. New Methods and Approaches for the Study of the Multiplicity of the Middle Ages in a Global Perspective (3rd 16th CE) (2014). http://www.academia.edu/8393940/Conflicts_as_a_structure-forming_force._The_reign_of_Henry_VII_1225-1235_in_network-analytic_perspective
7. Harary, F.: On the notion of balance of a signed graph. Mich. Math. J. **2** (1953). doi:10.1307/mmj/1028989917
8. Heider, F.: Attitudes and cognitive organization. J. Psychol. **21** (1946). doi:10.1080/00223980.1946.9917275
9. Reichardt, J., Bruggeman, J.: Statistical mechanics of community detection. Phys. Rev. E **74** (2006). doi:10.1103/PhysRevE.74.016110
10. Traag, V.A., Bruggeman, J.: Community detection in networks with positive and negative links. Phys. Rev. E **80** (2009). doi:10.1103/PhysRevE.80.036115

Comparative Topological Signatures of Growing Collaboration Networks

Siddharth Pal, Terrence J. Moore, Ram Ramanathan and Ananthram Swami

Abstract We study topological signatures in growing collaboration networks using standard and persistent homology. Persistent homology has thus far been primarily used for topological data analysis using a point cloud representation. In contrast, we apply persistent homology on temporal networks, and use it as a tool to compare and contrast between different growing networks. Specifically, we consider two collaboration networks: the paper collaboration network DBLP, and the actor collaboration network IMDB. We compare the evolution of their network properties, and of the homology (Betti numbers) with time. We also compare their topological signatures using persistent homology. We introduce a distance metric for comparing the topological signatures, and using it, visualize the similarity between individual segments through multidimensional scaling. We observe that, while the DBLP network has substantially evolved over time, the nature of collaboration in the IMDB network has relatively remained unchanged over the period 1950–2008. Our work shows that homology-based signatures can be effective in discriminating between real-world networks.

S. Pal (✉) · R. Ramanathan
Raytheon BBN Technologies, Cambridge, USA
e-mail: spal@bbn.com

R. Ramanathan
e-mail: ramanath@bbn.com

T.J. Moore · A. Swami
U.S. Army Research Lab, Adelphi, USA
e-mail: terrence.j.moore.civ@mail.mil

A. Swami
e-mail: ananthram.swami.civ@mail.mil

B. Gonçalves et al. (eds.), *Complex Networks VIII*,
Springer Proceedings in Complexity, DOI 10.1007/978-3-319-54241-6_18

1 Introduction

Networks of collaborations have received significant attention in the network science literature [11]. These networks are usually represented through graphs, where an edge exists between two vertices if the individuals corresponding to those vertices were part of a team or collaborative project. However, this approach obscures the fact that teams are a collection of individuals, collections often of size greater than two. Bipartite graphs (as affiliation networks), hypergraphs, and simplicial complexes are several of the proposed structural models often used to represent group structure in networks. Of these, we are interested in the study of collaborations modeled as a simplicial complex, studied in the social sciences at least as far back as [1]. This choice of structural representation enables the study of the algebraic topology of a set of collaborations using persistent homology.

Persistent homology is the primary tool in the growing field of topological data analysis [3, 6]. This computational topology method characterizes the homological structure of data over the range of a proximity scale parameter, where a monotonic sequence of parametric values creates a filtration of simplicial complexes that distinguishes persistent topological features of the data from noisy, or short-lived, features. Typically, the data is a point cloud and a set of points are connected as vertices of a simplex in a simplicial complex when the distance between pairs of points is below a distance threshold. It has been applied in a vast array of domains such as sensor networks [5], brain networks [8, 13], complex networks [7], etc. The topological persistence of collaboration networks has been previously studied where the proximity parameter is derived from the edge weights [4] (e.g., number of co-authored papers) and a notion of "research distance" [2].

In this work, we study the growing collaboration network with a *temporal parametrization*, i.e., we view the network over time and characterize the temporal changes in its topological features. The scenario is very different from the typical filtrations using a distance measure between node pairs over a point cloud. In the point cloud case, there exists a threshold distance above which every point in the cloud is pairwise connected. As was noted in the weight-based filtration [4] and is true for the temporal filtration, many homology classes representing cycles of collaborations persist indefinitely. This prohibits network comparison using the traditional topological data analysis approaches, e.g., Wasserstein and bottleneck distance. Carstens and Horadam [4] showed that appearance of nontrivial homology classes in real networks was unexpected compared with what would be expected in a random clique complex. Our approach is to use the barcode information [3] to construct a probability mass function (pmf) that models the relative homological growth of cycles and compare the pmfs.

The rest of the paper is organized as follows. In Sect. 2, we describe the simplicial complex model and temporal filtration. In Sect. 3, we detail our analysis on two classes of collaborations, namely scientific and movie actor collaborations. We conclude in Sect. 4 with a discussion of potential future work.

2 Temporal Persistent Homology of Collaboration Networks

A collaboration network is a network of individuals collaborating towards certain shared objectives. For example, the DBLP (Digital Bibliography & Library Project) computer science bibliography network is the network of authors who collaborated towards publishing a paper in fields related to computer science, and the IMDB (Internet Movie Database) collaboration network documents actors who worked together on a movie. While a collaboration network can be represented as a graph, it can also be represented as a mathematical object called simplicial complex.

Simplicial Complexes: A simplicial complex [10] is a pair $K = (V, S)$, where V is a finite set, and S is a set of non-empty subsets of V closed under the subset operation, i.e., for any $\rho \in S$ and $\tau \subset \rho$, we must have $\tau \in S$. Any set σ which belongs to the simplicial complex is called a simplex or a face. The dimension of a simplex is one less than the number of vertices in it, and the dimension of the simplicial complex is the maximum dimension among all the simplices.

We use Betti numbers to capture statistical properties of the topological space. Intuitively, the kth Betti number β_k is the number of k-dimensional surfaces which are unconnected via higher dimensions. Specifically, β_0 is the number of connected components, and β_1 is the number of 1-dimensional homology groups or 2-dimensional "holes", also referred to as cycles.

A collaboration network $\mathcal{N}$ is a pair (V, S), where V is a set of nodes, and $S \subseteq 2^V$ is the set of all collaborations in the network $\mathcal{N}$. Observe that a collaboration is closed under the subset operation, that is, for any collaboration $\rho \in S$ and $\tau \subset \rho$, we have $\tau \in S$ and hence a collaboration network can be captured by a simplicial complex. We represent each person in a paper or a movie as a vertex, and each collaborative act (and each of its subsets) as a simplex of vertices comprising it. Thus, for example, in the DBLP complex, each vertex represents a researcher and each simplex represents a collaboration relationship among the researchers on one or more papers.

Temporal Persistent Homology: Persistent homology is a method of obtaining a summary of the homological information of a topological space. As a first step, a filtration of simplicial complexes $\mathcal{T}$ needs to be constructed, where $\mathcal{T} = \{K_0, K_1, \ldots, K_n\}$ and $K_i \subseteq K_j$ for $i < j$, i.e., $V_i \subseteq V_j$ and $S_i \subseteq S_j$. In [4, 13], an unweighted graph is first obtained from a static weighted graph by removing all edges with weights lower than a threshold, which is then converted into a clique complex where all simplices or faces are cliques. Other works have considered point cloud data with a distance measure between node pairs [8]. For a particular threshold, a clique complex is constructed out of the point cloud data, where every pair of points in the complex is at a distance less than the threshold. A decreasing sequence of thresholds for the weighted graphs or an increasing sequence of thresholds for the point cloud leads to a filtration. Such filtrations are called weighted filtrations because they are parameterized by edge weights or distance thresholds.

This is fundamentally different from our approach where we construct a filtration from growing collaboration networks, where new collaborations are added each year.

Note that collaborations which have new nodes will lead to those nodes being added to the network, e.g., a new author or a new movie actor. Therefore, we have a sequence of networks $\{\mathscr{N}_t, t = 0, 1, \ldots, T\}$, where the network $\mathscr{N}_t$ represents the collaborations that occurred until time t. This sequence of collaborations constitutes a temporal filtration, where each complex in the filtration contains the previous complexes, i.e., $\mathscr{N}_t \subseteq \mathscr{N}_{t+1}$ ($V_t \subseteq V_{t+1}$ and $S_t \subseteq S_{t+1}$). Observe that in our approach, the data being used is inherently in the form of the relations defining the simplices and, hence, a simplicial complex, so no intermediate step of constructing a clique complex is required.

Once the filtration is obtained, the homology information is computed as described in [6]. This gives the details on the homology groups, which are the representative holes and higher dimensional voids, how long they persist, and their structure. The birth and death of these homology groups can be visualized through barcodes or persistence diagrams [6].

3 Experimental Results

In this section, we present experimental results on persistent homology of growing collaboration networks as defined in the previous section. We investigate if different types of real-world collaboration networks can be distinguished based on their topological signatures.

We study topological signatures in two different ways. First, in Sect. 3.1 we compare non-persistent homological and other properties on 10 year segments of IMDB and DBLP, to look for trends and insights. Next, in Sect. 3.2, we use persistent homology to compute distances between 10 year segments of DBLP and IMDB in order to investigate and quantify similarities and dissimilarities in their signatures. JavaPlex [14] was used to compute homology in both approaches.

3.1 Evolution of Homological Properties over Time

We consider 10 year windows in both datasets, with the median year of the windows taken every 5 years. We study the topological properties of networks corresponding to the 10 year windows, and investigate how they change with the median year of the window. We denote the non-overlapping DBLP network segments as follows: D_1 represents the segment 1950–59, D_2 represents 1960–69, and so on, until D_5 representing 1990–99 and finally D_6 representing 1999–2008. Similarly, I_1 represents the IMDB segment 1900–09, I_2 represents 1910–09, and so on, until I_{10} representing 1990–99 and finally I_{11} representing 1999–2008.

Our analysis indicates that the later DBLP network segments D_4, D_5 and D_6 have more than 99% of all holes in their largest connected component (LCC). Among the earlier DBLP segments, D_1 has only one hole which is in the LCC; D_2 has 6

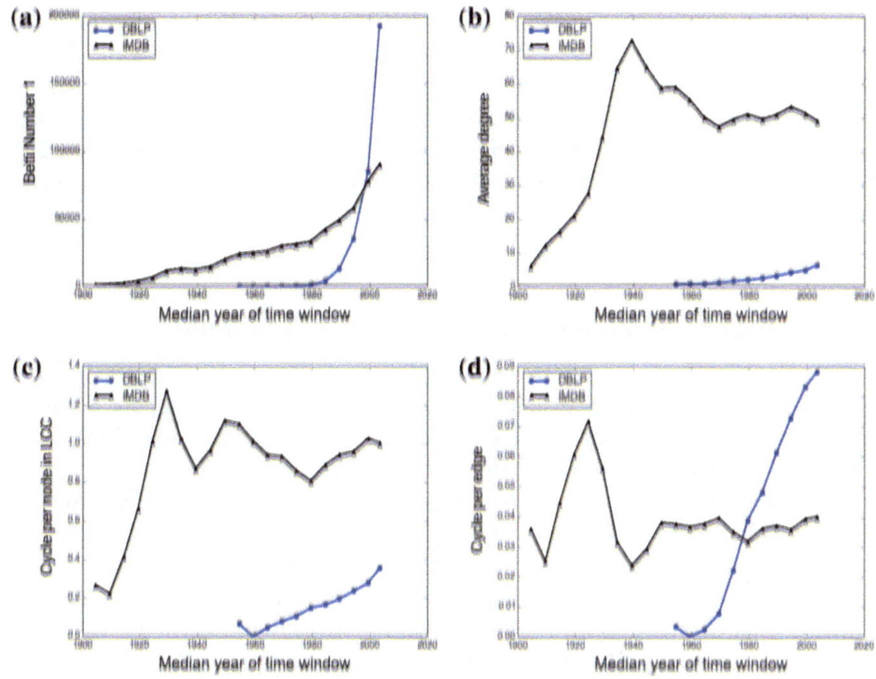

Fig. 1 Temporal evolution of network properties of IMDB and DBLP

holes, 4 of which are in the LCC, and D_3 has 358 holes out of which 346 (96.65% of total holes) are in the LCC. Other than I_2 which has 97.5% of all holes in the LCC, all the IMDB network segments have more than 99% of all holes in the LCC. Therefore, a significant percentage of holes is in the LCC for both the IMDB and DBLP networks. This empirical observation recalls the theoretical result in [12, p. 410] for Erdős-Rényi graphs, which states that the probability of small components being acyclic tends to one in the limit of large graph size, while noting the fact that Erdős-Rényi graphs are not good models for the IMDB/DBLP datasets.

In Fig. 1a, we see that the number of cycles grows steeply with time for the DBLP network especially starting from year 1990, whereas in the IMDB network the growth is more gradual, which eventually picks up around year 1980. Similarly, Fig. 1b indicates that the average degree has been falling in the IMDB network since year 1940, while having increased steadily for the DBLP network during the period 1970–2008. This indicates that authors in the DBLP network are working with greater number of authors over time as compared to actors in the IMDB network. From Fig. 1c it is evident that the number of cycles per node steadily increases with time for the DBLP network, while being largely unchanged for the IMDB network during the period 1930–1999. Furthermore, from Fig. 1d we observe that the number of cycles per edge increases with time for the DBLP dataset in the period 1970–1999, while it remains largely constant for the IMDB dataset for the period 1950–2008.

This analysis allows us to conclude that the density of cycles, both number of cycles per node and per edge increases with time for the DBLP dataset, while for the IMDB dataset it fluctuates or remains largely constant. We therefore observe that the homological properties of both networks change over time at different rates, which leads to the question whether they can be used to distinguish temporal networks. This thread will be taken up in the next subsection, which deals with finding distances between growing networks.

3.2 Persistent Homology Based Distance Computation

In the literature, bottleneck distance [6, 8] has been used to compute distances between topological signatures like persistence diagrams or barcodes. However, in growing network datasets we have infinitely-persistent nontrivial homology cycles as seen from the barcodes for the two networks (Fig. 2). This is different from the standard depiction of barcodes [7, 13] because almost all of the cycles never die out. This leads to a problem with the bottleneck distance or other similar distance measures on the persistence diagrams or barcodes, because the bottleneck distance is always infinity if there is a different number of infinitely-persistent cycles in each network. This is not a problem with filtrations based on point clouds with a distance parameter because at some distance threshold all points are connected and there is no higher-dimensional homology in the clique complex. We can cap the length of cycle lifetimes, to say $T + 1$ for a temporal filtration that ends at time T, but any bottleneck distance will still be heavily biased toward the difference between the number of persistent cycles in the two datasets. And that difference is influenced by the size of the networks, and in a small way by the clustering of the networks.

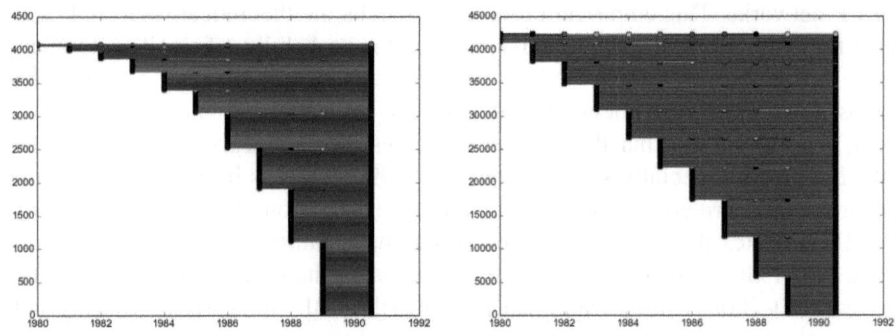

(a) DBLP (b) IMDB

Fig. 2 Barcodes for 1-dimensional homology groups or holes corresponding to period 1980–89 (Horizontal axis represents time in years, and each horizontal line represents a hole starting from its year of birth, continuing until it dies. Most of the holes persist beyond year 1990.)

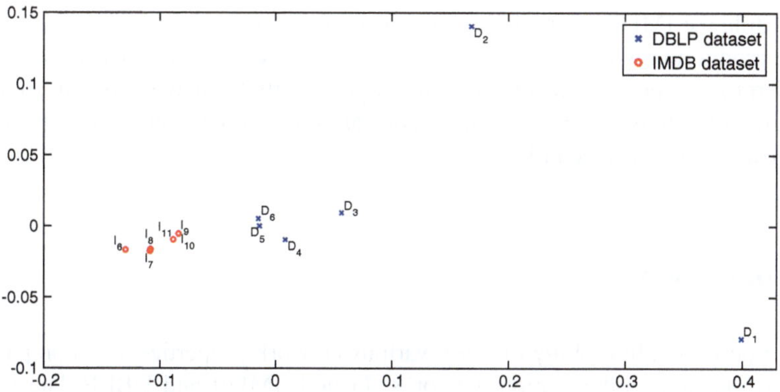

Fig. 3 Multi-dimensional scaling of the Jensen-Shannon distances

Since, bottleneck distance or other related distance measures cannot be used, we introduce a new distance measure between growing networks. This measure captures the difference in the rate of growth of cycles in the networks being compared.

Defining Distance Measures: For a sequence of growing networks $\{\mathcal{N}_t\ ,\ t = 0, 1, \ldots, T\}$, we define a function $g : \{0, 1, \ldots, T\} \rightarrow \mathbb{Z}^+$, such that $g(i)$ represents the number of holes newly formed at time i. These values can be obtained from Javaplex using the information on persistence intervals. From the function g, we define a cumulative distribution function as follows

$$F(x) = \frac{\sum_{t=0}^{x} g(t)}{\sum_{s=0}^{T} g(s)},\ x = 0, 1, \ldots, T, \tag{1}$$

and the corresponding pmf f. We shall use the Jensen-Shannon divergence [9] to compare the cycle growth rate of two networks. We only consider the birth times of holes because as argued previously most of the holes are infinitely-persistent. The Jensen-Shannon divergence between network segments are visually represented through multi-dimensional scaling in Fig. 3.

From Fig. 3 note that the IMDB segments are closely clustered together separately from all the DBLP segments. The DBLP segments are relatively more scattered with segments D_3 through D_6 clustered tighter, and the D_1 and D_2 segments being outliers. This leads us to conclude that while the IMDB dataset is topologically more self-similar over the period 1950–2008, the DBLP network changes over time, with the later years 1970–2008 being vastly different from the early years 1950–1969. This can be explained by going back to Sect. 3.1 which studies the evolution of homological properties of the networks. Observe that while the average degree (Fig. 1b), cycle per node (Fig. 1c), and cycle per edge (Fig. 1d) remain relatively constant over time for the IMDB network over the observed period, they increase substantially for the DBLP network especially starting from around year 1970. This suggests that the DBLP network changes significantly during the duration 1950–

2008 while the IMDB network does not. This could be attributed to the fact that the movie production industry being much older, was already mature by the year 1950, and therefore its properties did not change significantly from then onwards; whereas, the nature of collaboration has evolved substantially in computer science as the field has grown over time (see [15]).

4 Conclusion

We use persistent homology to study various network properties, and compare and contrast different collaboration networks. In both IMDB and DBLP datasets, the number of cycles grows with time, albeit at a different rate, while being predominantly restricted to the largest connected component. We study and compare the growth in cyclicity of the networks, with respect to time, and size of the LCC. We argue that existing distance measures between persistence barcodes does not seem to apply to temporal persistent homology, which leads us to develop a new distance metric for comparing evolving networks. Using this measure, we compare different segments of the DBLP and IMDB datasets, and conclude that the IMDB dataset is more self-similar than the DBLP dataset over the observed period 1950–2008. We also observe that the intra IMDB and intra DBLP distances are smaller than the inter IMDB-DBLP distances (barring early DBLP segments). From our study, it appears that topological signatures could be effective in discriminating between different temporal collaboration structures.

Acknowledgements Research was sponsored by the Army Research Laboratory and was accomplished under Cooperative Agreement Numbers W911NF-09-2-0053 (the ARL Network Science CTA). The views and conclusions contained in this document are those of the authors and should not be interpreted as representing the official policies, either expressed or implied, of the Army Research Laboratory or the U.S. Government. The U.S. Government is authorized to reproduce and distribute reprints for Government purposes notwithstanding any copyright notation here on. This document does not contain technology or technical data controlled under either the U.S. International Traffic in Arms Regulations or the U.S. Export Administration Regulations.

References

1. Atkin, R.H.: From cohomology in physics to q-connectivity in social science. Int. J. Man-Mach. Stud. **4**(2), 139–167 (1972)
2. Bampasidou, M., Gentimis, T.: Modeling collaborations with persistent homology (2014). arXiv:1403.5346
3. Carlsson, G.: Topology and data. Bulletin AMS **46**(2), 255–308 (2009)
4. Carstens, C.J., Horadam, K.J.: Persistent homology of collaboration networks. Math. Probl. Eng. (2013)
5. De Silva, V., Ghrist, R.: Coverage in sensor networks via persistent homology. Algebr. Geom. Topol. **7**(1), 339–358 (2007)

6. Edelsbrunner, H., Harer, J.: Persistent homology-a survey. Contemp. Math. **453**, 257–282 (2008)
7. Horak, D., Maletic, S., Rajkovic, M.: Persistent homology of complex networks. J. Stat. Mech. Theory Exp. **2009**(03), 3–34 (2009)
8. Lee, H., et al.: Persistent brain network homology from the perspective of dendrogram. IEEE Trans. Med. Imaging **31**(12), 2267–2277 (2012)
9. Lin, J., Wong, S.K.M.: A new directed divergence measure and its characterization. Int. J. General Syst. **17**(1), 73–81 (1990)
10. Munkres, J.R.: Elements of Algebraic Topology. Addison-Wesley (1984)
11. Newman, M.E.J.: The structure of scientific collaboration networks. Proc. Natl. Acad. Sci. **98**(2), 404–409 (2001)
12. Newman, M.: Networks: an Introduction. Oxford University Press (2010)
13. Petri, G., et al.: Homological scaffolds of brain functional networks. J. R. Soc. Interface **11**(101), 20140873 (2014)
14. Tausz, A., Vejdemo-Johansson, M., Adams, H.: Javaplex: a research software package for persistent (co) homology. Software available at http://code.google.com/javaplex. (2011)
15. Wuchty, S., Jones, B.F., Uzzi, B.: The increasing dominance of teams in production of knowledge. Science **316**(5827), 1036–1039 (2007)

Part VI
Human Behavior

Explaining Changes in Physical Activity Through a Computational Model of Social Contagion

Julia S. Mollee, Eric F.M. Araújo, Adnan Manzoor, Aart T. van Halteren and Michel C.A. Klein

Abstract Social processes play a key role in health behaviour. Understanding the underlying mechanisms of such processes is important when designing health interventions with a social component. In this work, we apply a computational model of social contagion to a data set of 2,472 users of a physical activity promotion program. We compare this model's predictions to the predictions of a simple linear model that has been derived by a regression analysis. The results show that the social contagion model performs better at describing the pattern seen in the empirical data than the linear model, indicating that some of the dynamics of the physical activity levels in the network can be explained by social contagion processes.

Keywords Computational modelling · Social contagion · Validation · Physical activity

Research was sponsored by the Army Research Laboratory and was accomplished under Cooperative Agreement Numbers W911NF-09-2-0053 (the ARL Network Science CTA). The views and conclusions contained in this document are those of the authors and should not be interpreted as representing the official policies, either expressed or implied, of the Army Research Laboratory or the U.S. Government. The U.S. Government is authorized to reproduce and distribute reprints for Government purposes notwithstanding any copyright notation here on. This document does not contain technology or technical data controlled under either the U.S. International Traffic in Arms Regulations or the U.S. Export Administration Regulations.

J.S. Mollee · E.F.M. Araújo · A. Manzoor · A.T. van Halteren · M.C.A. Klein (✉)
VU University Amsterdam, Amsterdam, The Netherlands
e-mail: m.c.a.klein@vu.nl

J.S. Mollee
e-mail: j.s.mollee@vu.nl

E.F.M. Araújo
e-mail: e.araujo@vu.nl

A. Manzoor
e-mail: a.manzoorrajper@vu.nl

A.T. van Halteren
e-mail: a.t.van.halteren@vu.nl; aart.van.halteren@philips.com

A.T. van Halteren
Philips Research, Eindhoven, The Netherlands

B. Gonçalves et al. (eds.), *Complex Networks VIII*,
Springer Proceedings in Complexity, DOI 10.1007/978-3-319-54241-6_19

1 Introduction

Physical inactivity is a major worldwide concern, as it can lead to many long-term health risks [6, 7]. These risks can be reduced if an adult fulfills the requirement (according to recommendations of the WHO and other public health organizations) of at least 150 min of moderate or 75 min of vigorous intensity physical activity per week, or a combination of both [8, 19]. An active lifestyle not only improves a person's physical health, but it also has positive effects on mental health [13].

If used in innovative ways, eHealth and mHealth hold great potential to steer physical activity promotion programs in the right direction and let greater numbers of people benefit from it. However, this requires the right choices about the way in which technology is embedded in these programs. For example, simply using a wearable device alone will not suffice to achieve sustainable behaviour change [14]. To maintain new behaviour for a longer period of time, other important ingredients are needed, e.g. evidence-based techniques such as goal setting and timely feedback, and a supportive social environment.

Social processes play a key role in health behaviour. It has been shown that people become more successful in maintaining a healthy lifestyle when they function within their social context [18, 20]. In addition, the social environment enables people to compare their physical activity achievements with their peers or to seek social support from them. Within online social networks, this is commonly implemented via leader boards with achievements, building on the theory of social comparison [17]. Overall, in the context of health promotion programs, social processes can provide a leveraging mechanism to achieve and maintain a healthy lifestyle. Understanding these mechanisms is therefore important.

In this paper, we use a data set about health behaviour in a social context to understand the underlying social processes. It is a continuation of earlier work on this subject [10, 11]. In [11], a large data set of an online physical activity promotion program was used to compare the physical activity levels of people who are part of an online social network with those who did not opt to join the network. One of the conclusions was that participants who are part of an online community have significantly higher activity levels and a higher increase in activity compared to participants who chose not to become part of the community. However, this did not answer the question what kind of social phenomenon was causing the higher activity levels.

In this work, we try to answer the question whether the increase in physical activity can be explained by social contagion [5]. Our main hypothesis is that the higher activity levels of the community users can be partially explained by social contagion and partially by the effect of the health promotion program. The research question is addressed by comparing the activity data of the participants with two types of predictions: (1) based on a simple linear model that captures the effect of participating in the program and the online community, and (2) based on a model of social contagion combined with the linear model.

2 Background

Because a majority of the adults in the Western world does not meet the guidelines for physical activity, public health professionals are aiming at population-wide interventions. Since decades, the area of preventive medicine is investigating how people can be stimulated to be more physically active [15]. More recently, the smartphone has been discovered as tool for measuring and influencing physical activity [3]. Many of these technology-mediated interventions use some kind of social influence. A specific appearance of social influence is the phenomenon of social contagion [5]. It has been shown that people can influence each other via their social networks up to three degrees of distance. Although these claims have been criticized [16], one could imagine that people transitively influence each other via social relations.

In [2, 4], a temporal-causal computational model is presented that describes how the mutual absorption of emotions in a social network affects the emotions of the individuals. This model was used for the study that is described in this paper. Our assumption is that physical activity behaviour is influenced by internal states like motivation, attitudes and goals, and that those spread in a similar way as described in the model of emotion contagion.

The model [2] describes how internal state q_A of person A affects the internal states of other persons B_i. This process is determined by the strength by which the state is *expressed* (ε_A), the *openness* of the receiver (δ_B) and the strength of the channel between them (α_{AB}). Together, these factors determine the *connection weight* ω_{AB}. Thus, the impact $\mathbf{impact}_{AB}(t)$ of the state of person A on the state of person B is:

$$\mathbf{impact}_{AB}(t) = \omega_{AB} q_A \tag{1}$$

The aggregated impact $\mathbf{aggimpact}_B(t)$ at time t of the states q_{A_i} of all connected persons on state q_B is modelled as a scaled sum. From this it follows that $\mathbf{aggimpact}_B(t)$ is calculated as a weighted average of all the impacts of the different connections of a person:

$$\mathbf{aggimpact}_B(t) = \sum_{A_i \neq B} w_{AB} q_{A_i}(t) \tag{2}$$

with w_{AB} chosen in such a way that it is proportional to ω_{AB} and the sum of all weights is 1. The new state for each person in the network is calculated by integrating some factor η of the aggregated impact:

$$contagion\_effect(t) = \eta_A[\mathbf{aggimpact}_B(t) - q_B(t)] \tag{3}$$

$$q_B(t + \Delta t) = q_B(t) + contagion\_effect(t)\Delta t \tag{4}$$

For the purpose of this study, we assumed that all people have the same expressiveness and openness, and that all connections were of the same strength. This was done out of necessity, as our data set does not contain specific information about these factors. The model's parameters for openness, expressiveness and channel strength were thus set to a default value of 0.5.

3 Methods

This section describes how the data was collected and preprocessed, as well as what types of analyses were run.

3.1 Data Collection

The data originates from a physical activity promotion program in which participants are asked to wear an activity monitor that measures physical activity level (PAL) using an accelerometer. Based on the activity data that is repeatedly uploaded by the participants, the program stimulates them towards a more active lifestyle by gradually increasing the weekly activity targets over a 12-week activity plan. The baseline for this activity plan is established in an initial assessment week. After completing a plan, participants can choose to take another 12-week activity plan or decide to remain at the level of their last completed plan.

After the initial assessment week, participants also get access to a dashboard with information about energy expenditure (calories burnt) and their achievements relative to a weekly goal. The program provides an opt-in online community that allows participants to establish connections and to compare achievements. Each participant in the community will see how their achievements rank compared to other participants with whom they are connected. Community participants see the ranking within their own network each time they upload data from their activity monitor. The network structure and some social network analyses are discussed in [1].

3.2 Data Preprocessing

The original data set contains data for 52,788 users. Since the aim of this paper is to demonstrate the influence of social contagion on people's physical activity levels, we are only interested in the 5,041 users who opted in for the online community of the program.

First, any participants that joined the program for testing purposes or users with missing information, such as gender or body mass index (BMI), were removed from the data set, as well as participants that didn't have a start date for their first plan. The resulting data set contains participants for whom valid physical activity data is available. The network was further pruned by removing connections that were initiated by one participant, but never confirmed by the other participant.

As the online community feature was not part of the program until April 28th 2010, all data before that date was disregarded. Community data was available until August 6th 2010, but the PAL data was incomplete for the last couple of days. This can be explained by the fact that some users did not upload their data for those days yet. Therefore, only the data up to July 28th 2010 was considered, resulting in a data selection that spanned a period of 91 days.

Within this period of 91 days, only active and connected participants were included in the current analysis. In other words, any users who entered the program, but did not join the online community, or users that dropped out of the program before this period started, were removed from the data set. This data cleaning process leaves us with 2,472 relevant nodes in the period between April 28th 2010 and July 28th 2010.

Although the primary unit of physical activity in the data set is the PAL, users see percentages of their goal achieved rather than the PAL itself on their online dashboard. The ranking with connected users on is also based on this relative performance. Therefore, our analyses are also based on the ratios of goals achieved, i.e. the current PAL divided over the target PAL.

3.3 Model Simulations

Previous work has shown that the combination of participating in the program and joining the online community is associated with a small but significant average increase in PAL [11]. The objective of the current work was to demonstrate whether the dynamics of users' physical activity levels can be (partially) explained by social contagion. Therefore, we compared the predictive performance of two different models: (1) a simple linear model, that describes the effect of the program on community members; and (2) a combined model, that captures the social contagion process and incorporates the known linear increase as well.

Scenario 1: Simple linear model.

The simple linear model describes the effect of the physical activity promotion program and the online community on the users' physical activity levels. Previous analyses have shown that this effect is an average PAL increase of 0.0005821 per day [11]. These analyses were based on a subset of users from the same data set, with all users being in their first plan and member of the community. The increase in PAL translates to an increase in energy expenditure of 1.05 kCal for an average male with a basal metabolic rate (BMR) of 1800 kCal/day [12].

To translate this increase in PAL to the unit predicted by the model (i.e., the goal achieved), the simple linear model adds a daily increase of 0.0005821 divided by the current target PAL to the user's goal achieved, as shown in Eqs. 5 and 6.

$$linear\_effect(t) = \frac{0.0005821}{target\_pal(t)} \quad (5)$$

$$goal\_achieved(t+\Delta t) = goal\_achieved(t) + linear\_effect(t) \quad (6)$$

Scenario 2: Combined social contagion model.

The combined social contagion model describes the linear increase in PAL as well, but combines it with the model of social contagion that captures the dynamics between the nodes in the network, as summarized in Eq. 7, where *contagion_effect(t)* denotes the social contagion effect as described in Sect. 2, Eq. 3. In this case, the state q represents the percentage of goal achieved. By enriching the social contagion model with the daily increase in PAL (as in the simple linear model), we account for the demonstrated stimulating effect of the program and the community, and thereby nullify a possible disadvantage on the social contagion model.

$$goal\_achieved(t+\Delta t) = goal\_achieved(t) + contagion\_effect(t) + linear\_effect(t) \quad (7)$$

As mentioned in Sect. 3.2, the analyses were based on the predictions of the goal achieved, i.e. the proportion of the target PAL achieved by the user, rather than the user's current PAL. Additionally, the model predictions were done for users in their first plan. Of the 2,472 relevant users identified in Sect. 3.2, 1,939 were participating in their first plan for at least part of the time period under consideration. The reason behind this choice is that users in their first plan are most comparable to the general population: they have just entered the program, and therefore have no prior knowledge of or experience with the plans or other parts of the intervention. Also, it is likely that people in their first plan have the highest adherence rates and interact more with the program, which makes them a more interesting population as well. However, users who have not yet started or already completed their first plan can still influence users in their first plan through social contagion. Therefore, they are considered by the social contagion model, but only as input of the contagion process towards the users under consideration (i.e., users in their first plan).

To run the models, the initial values have to be determined. For all users for whom a target PAL is not available (i.e., users who are in their assessment week and have yet to start their first plan), the initial goal achieved value was based on the average PAL of their assessment week and their first target PAL. For all users with a target PAL, the initial goal achieved was calculated by dividing the average PAL for one week before the start date of the simulations (i.e., April 28th 2010) by the current target PAL. If for some reason, no data was available for that week, the initial goal achieved was based on the average PAL in the month prior to the start date of the simulations.

In the social contagion model, we used the initial goal achieved values of the simulated nodes as described above, and the empirical data from the surrounding nodes as input to the contagion process. This choice was motivated by the fact that we were only interested in simulating the effect of the behaviour of users on users in their first plan, rather than simulating the behaviour of those other users as well.

3.4 Analyses

To evaluate the accuracy of the two models, we first calculated their average predictions for the approximately 1,939 users in their first plan in the data set, as well as the average goal achieved values based on the empirical data. Based on these values, we tested whether there is a significant difference in the magnitude of the errors of the two models with a Mann Whitney U test. In addition, we determined the correlations of both models' predictions to the empirical data by means of Mann Kendall tests.

4 Results

As explained in Sect. 3.2, after thorough preprocessing of the data, 2,472 relevant users remained in the period between April 28th and July 28th 2010.

Following the procedures described in Sect. 3.3, the two models were run on the initial data. Figure 1 provides an impression of the predicted goal achieved values for the 1,939 users in their first plan by the two models. The simulation of the linear model shows a steady increase in the goal achieved. The combined model shows the effect of the contagion between the users, in combination with the steady increase. Any interruptions of the lines in either plot are caused by users entering the program or community, or by users dropping out of the program.

After averaging the model predictions, as well as the empirical data, for all users in their first plan per day, the graph in Fig. 2 was obtained. It shows the average predictions of the linear model (green) and the combined model (blue), and the

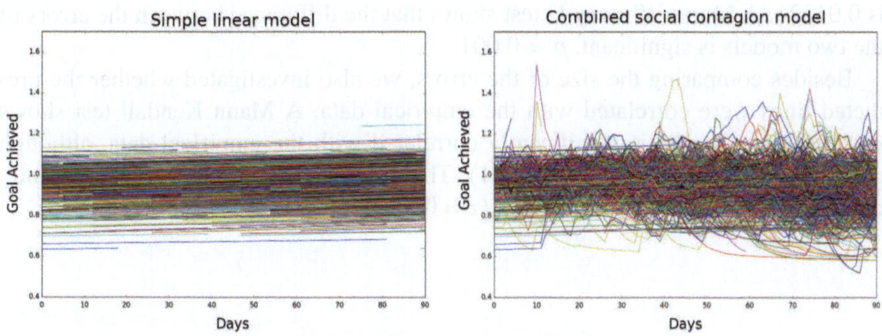

Fig. 1 Predictions of the simple linear model (*left*) and the combined model (*right*)

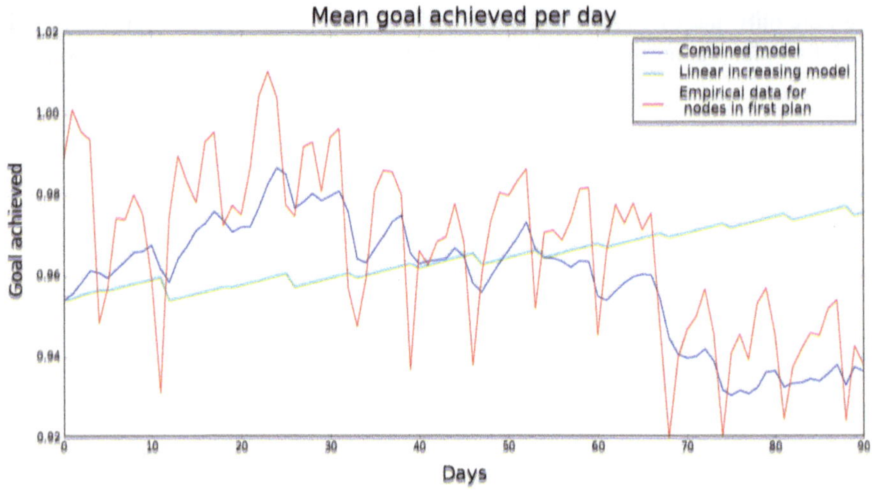

Fig. 2 Average predictions of the two models (*green* linear, *blue* combined), and the empirical data (*red*)

Table 1 Model evaluations

	Absolute error		Kendall's correlation test	
	Mean	St. Dev.	Kendall's τ	Kendall's p
Linear model	0.02212	0.01378	–0.46227	<0.001
Combined model	0.01321	0.00855	0.53895	<0.001

empirical data (red). The sharp troughs in the empirical data mark the Sundays, when physical activity levels on average are substantially lower.

Figure 2 already gives the impression that the combined model is much closer to the empirical data than the linear model. Indeed, the mean absolute error (MAE) of the linear model is 0.02212, whereas the mean absolute error of the combined model is 0.01321. A Mann-Whitney U test shows that the difference between the errors of the two models is significant, $p < 0.001$.

Besides comparing the size of the errors, we also investigated whether the predicted lines were correlated with the empirical data. A Mann Kendall test shows that the linear model is significantly correlated with the empirical data, although negatively ($\tau = -0.46227$, $p < 0.001$). The combined model is also significantly correlated, but in this case positively ($\tau = 0.53895$, $p < 0.001$) (Table 1).

5 Conclusions

The results described in Sect. 4 show that the combined model, which integrates the social contagion model with a steady linear increase in PAL, is indeed better able to capture the dynamics of the physical activity levels in our data set than the linear model. Its predictions show a significant positive correlation with the empirical data. Additionally, the errors of the combined model's predictions are significantly smaller than those of the linear model.

One of the main strengths of this work is its foundation on a large set of empirical data covering several months. Careful and extensive preprocessing of the empirical data was conducted to ensure data that is sensible for the simulated models. For example, we dynamically removed connections to users who practically dropped out of the program (but were still in the system), to prevent their (missing) data from affecting the results.

Another strength of our work is that we compared the performance of the model we were mainly interested in to an *informed* linear model. That way, we do not impose a disadvantage on the baseline model, thus increasing the chances of superiority of our more complex model. However, it is interesting to see that the empirical data shows a development that is actually opposite to the direction of the linear increase model. One possible explanation for this observation could be that the linear increase was found after aligning the data by the day in the program rather than the calendar date. The pattern in the current data set is then caused by users in different phases of the first plan entering and leaving the program over time (e.g., because their first plan is finished halfway the period that we selected). A second possible explanation is that the linear model describes an increase in PAL, whereas it is transformed and applied to the *progress towards the target PAL* in this work. A third possible explanation is that the linear model was based on a different subset of the same data set, so maybe the subset analysed in this work does not show an average increase in PAL.

One of the limitations of this work is its restricted generalizability. As all analyses were based on data collected in the context of a physical activity promotion program (see also Sect. 3.1), the results cannot directly be transferred to the general population. However, by choosing to focus on people who are exposed to the program for the first time, we have tried to minimize that discrepancy.

Another limitation is that the social contagion model only considers the online community as the network through which the behaviour spreads, although contagion also takes place on different levels and in different contexts. Additionally, we did not take into account whose data is actually shown on the user's dashboard: all connections were treated equally, whereas the performance of friends may not be shown on the dashboard when the difference was too big (e.g., more than 10 position difference). Future work could reveal whether limiting the contagion model to only the connected users who are visible on the dashboard improves the performance of the model. A further limitation is that we used default values of 0.5 for the parameters (for expressiveness, channel strength and openness) in the combined model. In future work, we could investigate whether using calibrated values would yield better results.

It is also possible to experiment with models that incorporate the principle of non-linearity in behaviour change, e.g. by exploiting thresholds for effects [9].

Up to our knowledge, we present the first analysis of the ability of a computational model of social contagion to capture the pattern of physical activity levels in a community over time. The results show that the enriched social contagion model performs better at describing the pattern in the empirical data than the linear model, indicating that some of the dynamics of the physical activity levels in the network can be explained by social contagion processes. This is vital information for designers of health interventions with a social component, as such models can then be used to maximize the benefits of social influence processes.

References

1. Araújo, E.F.M., Klein, M.C.A., van Halteren, A.T.: Social connection dynamics in a health promotion network. In: Complex Networks 2016—The 5th International Workshop on Complex Networks and their Applications (2016)
2. Araújo, E.F.M., Treur, J.: Analysis and Refinement of a Temporal-Causal Network Model for Absorption of Emotions, pp. 27–39. Springer (2016)
3. Bort-Roig, J., Gilson, N.D., Puig-Ribera, A., Contreras, R.S., Trost, S.G.: Measuring and influencing physical activity with smartphone technology: a systematic review. Sports Med. **44**(5), 671–686 (2014)
4. Bosse, T., Duell, R., Memon, Z.A., Treur, J., van der Wal, C.N.: Agent-based modeling of emotion contagion in groups. Cogn. Comput. **7**(1), 111–136 (2015)
5. Christakis, N.A., Fowler, J.H.: Social contagion theory: examining dynamic social networks and human behavior. Stat. Med. **32**(4), 556–577 (2013)
6. Conn, V.S., Hafdahl, A.R., Mehr, D.R.: Interventions to increase physical activity among healthy adults: meta-analysis of outcomes. Am. J. Public Health **101**(4), 751–758 (2011)
7. Eime, R.M., Young, J.A., Harvey, J.T., Charity, M.J., Payne, W.R., et al.: A systematic review of the psychological and social benefits of participation in sport for children and adolescents: informing development of a conceptual model of health through sport. Int. J. Behav. Nutr. Phys. Act. **10**(98), 1 (2013)
8. Garber, C.E., Blissmer, B., Deschenes, M.R., Franklin, B.A., Lamonte, M.J., Lee, I.M., Nieman, D.C., Swain, D.P.: American college of sports medicine position stand. quantity and quality of exercise for developing and maintaining cardiorespiratory, musculoskeletal, and neuromotor fitness in apparently healthy adults: guidance for prescribing exercise. Med. Sci. Sports Exerc. **43**(7), 1334–1359 (2011)
9. Giabbanelli, P.J., Alimadad, A., Dabbaghian, V., Finegood, D.T.: Modeling the influence of social networks and environment on energy balance and obesity. J. Comput. Sci. **3**(12), 17–27 (2012)
10. Groenewegen, M., Stoyanov, D., Deichmann, D., van Halteren, A.: Connecting with active people matters: the influence of an online community on physical activity behavior. In: International Conference on Social Informatics. pp. 96–109. Springer (2012)
11. Manzoor, A., Mollee, J.S., Araújo, E.F.M., Van Halteren, A.T., Klein, M.C.A.: Online sharing of physical activity: Does it accelerate the impact of a health promotion program? In: IEEE International Conference on Social Computing and Networking (SocialCom 2016). pp. 201–208. IEEE (2016)
12. Mifflin, M.D., St Jeor, S.T., Hill, L.A., Scott, B.J., Daugherty, S.A., Koh, Y.O.: A new predictive equation for resting energy expenditure in healthy individuals. Am. J. Clin. Nutr. **51**(2), 241–247 (1990)

13. Pate, R.R., Pratt, M., Blair, S.N., Haskell, W.L., Macera, C.A., Bouchard, C., Buchner, D., Ettinger, W., Heath, G.W., King, A.C., et al.: Physical activity and public health: a recommendation from the centers for disease control and prevention and the american college of sports medicine. Jama **273**(5), 402–407 (1995)
14. Patel, M.S., Asch, D.A., Volpp, K.G.: Wearable devices as facilitators, not drivers, of health behavior change. Jama **313**(5), 459–460 (2015)
15. Sallis, J.F., Owen, N.: Physical Activity and Behavioral Medicine, vol. 3. SAGE Publications (1998)
16. Shalizi, C.R., Thomas, A.C.: Homophily and contagion are generically confounded in observational social network studies. Sociol. Methods Res. **40**(2), 211–239 (2011)
17. Suls, J.E., Wills, T.A.E.: Social Comparison: Contemporary Theory and Research. Lawrence Erlbaum Associates Inc (1991)
18. Wing, R.R., Jeffery, R.W.: Benefits of recruiting participants with friends and increasing social support for weight loss and maintenance. J. Consul. Clin. Psychol. **67**(1), 132 (1999)
19. World Health Organization: Global recommendations on physical activity for health (2010). http://www.who.int/dietphysicalactivity/publications/9789241599979/en/
20. Zimmerman, R.S., Connor, C.: Health promotion in context: the effects of significant others on health behavior change. Health Educ. Behav. **16**(1), 57–75 (1989)

Everyday the Same Picture: Popularity and Content Diversity

Alessandro Bessi, Fabiana Zollo, Michela Del Vicario, Antonio Scala, Fabio Petroni, Bruno Gonçcalves and Walter Quattrociocchi

Abstract Facebook is flooded by diverse and heterogeneous content, from kittens up to music and news, passing through satirical and funny stories. Each piece of that vivid production reflects the heterogeneity of the underlying social background and provides sometimes interesting opportunities for the study of social dynamics. Indeed, in Facebook we found an interesting case: a page having more than 40 K followers that every day posts the same picture of a popular Italian singer. We use such a peculiar page as a baseline for the study and modeling of the relationship between content heterogeneity and popularity. In particular, we perform a comparative analysis of information consumption patterns with respect to pages posting heterogeneous content (science and conspiracy news). We conclude the paper by introducing a model mimicking users selection preferences accounting for the heterogeneity of contents.

1 Introduction

Online social networks such as Facebook foster the aggregation of people around common interests, narratives, and worldviews. Indeed, the World Wide Web caused a paradigm shift in the production and consumption of contents that increased both

A. Bessi
Information Sciences Institute, University of Southern California, Los Angeles, CA, USA

F. Zollo · M. Del Vicario · W. Quattrociocchi
IMT Institute for Advanced Studies, Lucca, Italy

A. Scala
ISC CNR, Rome, Italy

F. Petroni
Sapienza University of Rome, Rome, Italy

B. Gonçcalves (✉)
Center for Data Science, New York University, New York, NY, USA
e-mail: bgoncalves@gmail.com

B. Gonçalves et al. (eds.), *Complex Networks VIII*,
Springer Proceedings in Complexity, DOI 10.1007/978-3-319-54241-6_20

its volume and heterogeneity. Users can express their attitudes by producing and consuming heterogeneous information—e.g. conspiracists avoid mainstream news and follow their own information sources, whereas debunkers try to inhibit the diffusion of false claims. Images of kittens and pets, political memes, gossip, scandals spread on Facebook. By liking, commenting, and sharing their preferred contents, users express their passions and emotions—with sarcasm being no exception. In particular, pages promoting parody and sarcastic imitations of online social dynamics are common occurrences—e.g., *Ebola and Kittens* [1] or *In favor of chem-trails* [2]—An interesting case in Facebook is a page [3] with more than 40 K followers that posts everyday the exactly alike picture of Toto Cutugno, a famous Italian pop-singer.

In this work, we use this page as a baseline with which to study the effect of content diversity on popularity/virality. Specifically, we analyze user activity and post consumption patterns on the baseline page for a timespan of about 4 months. Through a comparative analysis between two sets of pages producing heterogeneous contents, we show that there are no remarkable differences in user activity patterns, whereas significant dissimilarities between post consumption patterns emerge. Such a comparative analysis allows to model information consumption accounting for the heterogeneity of contents. Hence, we show that the proposed model is able to reproduce the phenomenon observed from empirical data. In particular, we show the effects of different levels of contents' heterogeneity on posts consumption patterns.

The remainder of the paper is structured as follows. Background and Related Work reviews the literature on the study of social dynamics in online social media, stressing the challenges raised by the economy of attention. In Data Description we describe the Facebook dataset we used, whereas in Preliminaries and Definitions we explain some of the statistical tools we use throughout the paper. In Results and Discussion we show some statistical signatures concerning user activity and post consumption patterns, and then we introduce and discuss our data-driven model of information consumption. Finally, Concluding Remarks summarizes our findings.

2 Background and Related Works

A large body of literature addresses the study of social dynamics on socio-technical systems from social contagion to social reinforcement [4, 9, 13–16, 20, 23–25, 30–37, 46]. Among these, one of the most defining topics of computational social science is the understanding of the driving forces behind content popularity [44]. This challenge is typically addressed by analyzing the sentiment of comments, post, and users' attention [7, 19, 22, 27, 28, 38, 42, 45, 49]. However, the mechanisms behind popularity remain largely unexplored [21, 29, 47]: Why do some pieces of content become viral while other, seemingly identical, languish in obscurity? In [40] the authors tackle this question experimentally by measuring the impact of content quality and social influence on the eventual popularity or success of cultural artifacts. The effects of specific contents on the formation of communities of interest, their permeability to false information, and the resistance to changes were recently characterized in [10–12, 39] while in [5] the authors observe that connectivity patterns of the Facebook social network are prominently driven by homophily of users—i.e., the tendency of individuals to associate with others that are similar to them—towards specific kinds of contents. Microblogging platforms such as Facebook and Twitter [43] have lowered the cost of information production and broadcasting, boosting the potential reach of each idea or meme [8, 17]. Still, the abundance of information to which we are exposed through online social networks and other socio-technical systems is rapidly exceeding our capacity to consume it [48] causing information dynamics to be attention driven more than it had ever been before [18, 26, 41]. We further this debate and study the interlink between content diversity and popularity.

3 Data Description

In this work, we aim at investigating the role of content diversity on the dynamics of information consumption in online social networks. To this end, we use a set of Facebook pages promoting heterogeneous contents and a Facebook page promoting always the same picture. The set of pages promoting heterogeneous contents is composed by 73 public Facebook pages, whereof 34 are about science news and 39 are about conspiracy theories; we refer to the former as *science pages* and to the latter as *conspiracy pages* [11]. Using two significantly different kinds of topics we are also able to control for topical and community variety since there is little overlap between the users of both groups of pages. To further ground this analysis we use a page promoting homogeneous contents. This page, "La stessa foto di Toto Cutugno ogni giorno" ("Everyday the same photo of Toto Cutugno") publishes exclusively the same picture of the Italian singer every day, making it the perfect baseline; we refer to this page as the *baseline page*. We collected all the *likes* and *comments* to every post in each page, as well as the number of *shares*. The dataset includes all activity in the science and conspiracy pages for the period between August 22, 2013 and December 31, 2013, as well as all activity for the baseline page between August 22, 2014 (when the page was created) and December 31, 2014. In total, we collected around 2 M likes and 190 K comments, made by about 340 K and 65 K users, respectively. In Table 1 we summarize the details of our dataset. Likes, shares, and comments have different semantic meanings: a 'like' is a positive feedback on the post; a 'share' expresses approval and the will to divulge it further; while a 'comment' is a form to participate in collective debate and can be both positive or negative.

4 Preliminaries and Definitions

Here we provide some of the basic definitions that we use throughout the overall paper.

Table 1 Dataset statistics. The number of pages, posts, likes, comments, shares, likers, and commenters for science pages, conspiracy pages, and the baseline page

	Total	Science	Conspiracy	Baseline
Pages	74	34	39	1
Posts	49,354	13,028	36,169	157
Likes	2,095,677	614,078	1,184,084	297,515
Comments	192,967	40,608	138,138	14,221
Shares	3,782,480	477,457	3,297,687	7,336
Likers	344,367	162,146	159,524	22,697
Commenters	64,903	18,358	41,666	4,875

Statistical Tools. The Probability Density Function (PDF) of a real–valued random variable is a function f_X that describes the probability of the random variable falling within a given range of values, so that

$$\mathbf{Pr}[a \leq X \leq b] = \int_a^b f_X(x)\,dx.$$

The cumulative distribution function (CDF) of a real–valued random variable X is defined as

$$F_X(x) = \mathbf{Pr}(X \leq x) = \int_{-\infty}^{x} f_X(u)\,du.$$

Similarly, the complementary cumulative distribution function (CCDF) is defined as one minus the CDF, so that

$$C_X(x) = 1 - F_X(x) = \mathbf{Pr}(X > x) = \int_x^{\infty} f_X(u)\,du.$$

Notice that in order to compare metrics related to pages showing different activity and consumption volumes, we perform the unity–based normalization to bring all values in the range [0, 1].

Bipartite Networks. In our model we consider a bipartite network having as nodes users and posts. A like to a given post determines a link between a user and a post. More formally, a bipartite graph is a triple $\mathcal{G} = (A, B, E)$ where $A = \{a_i \mid i = 1 \ldots n_A\}$ and $B = \{b_j \mid j = 1 \ldots n_B\}$ are two disjoint sets of vertices indicating, respectively, users and posts, and $E \subseteq A \times B$ is the set of edges—i.e. edges exist only between vertices of the two different sets A and B. The bipartite graph $\mathcal{G}$ is described by the matrix M defined as

$$M_{ij} = \begin{cases} 1 \; if\, an\, edge\, exists\, between\, a_i\, and\, b_j \\ 0 \qquad\qquad otherwise \end{cases}$$

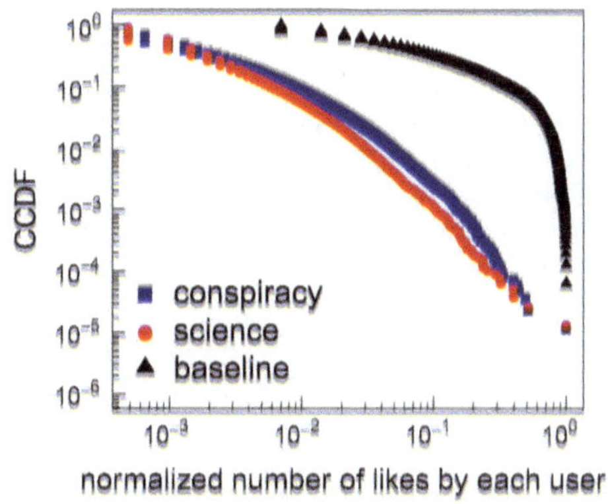

Fig. 1 Users' activity patterns. Complementary cumulative density function (CCDF) for the normalized number of likes by each user

Thus, $M_{ij} = 1$ means that a user $a_i \subset A$ liked a post $b_j \subset B$. It follows that the bipartite projection of users is a network of users in which a user $a_x \subset A$ is linked to a user $a_y \subset A$ if and only if both liked a given post $b_z \subset B$, i.e. if and only if

$$M_{xz} = 1 \wedge M_{yz} = 1.$$

5 Results and Discussion

In this section, we first present the statistical signatures characterizing users activity on pages with diversified content on specific topics (science and conspiracy news) against the case of the page posting every day the same picture (baseline). Then, we derive a model of information consumption mimicking user preferences with respect to contents.

5.1 *Content and Users Activity*

Let us focus on some regularities concerning users' activity on science pages and conspiracy pages compared with the baseline page. Figure 1 shows the complementary cumulative density function (CCDF) for the normalized[1] number of likes for each user.

In Fig. 2 we show the CCDF of the users' lifetime in terms of their liking activity—i.e. the temporal interval between the first and the last like of the user on a given page.

These figures show that users activity patterns are similar and present heavy-tailed distributions despite the different nature of the contents, and we can not find any

[1] We rescaled the number of likes to bring all values in the range [0, 1].

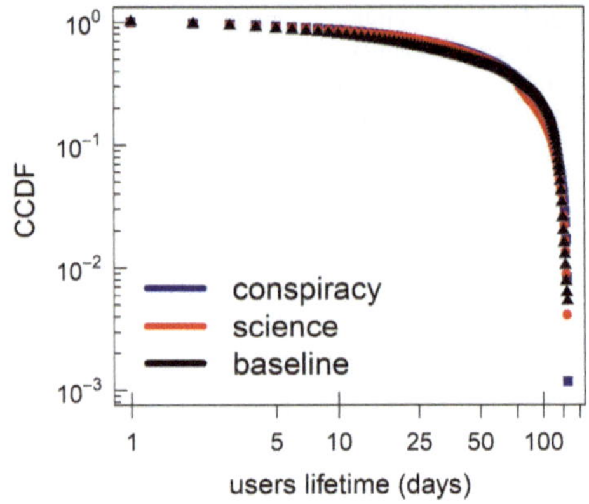

Fig. 2 Users' lifetime. Complementary cumulative density function (CCDF) of the users' lifetime in terms of their liking activity. The CCDF shows a slight difference in the lifetime of the baseline users with respect to science and conspiracy users

significant difference between the users interaction patterns induced by heterogeneous or homogeneous contents.

Conversely, by analyzing consumption patterns related to posts, we find a significant difference in the information consumption dynamics. Figure 3 shows the PDF for the number of likes received by posts belonging to science pages, conspiracy pages, and the baseline page. The number of likes received by posts are heavy–tailed distributed if the posts belong to pages promoting heterogeneous contents (science and conspiracy pages); whereas they are approximately distributed according to a Gaussian if the posts belong to a page promoting homogeneous content (baseline page).

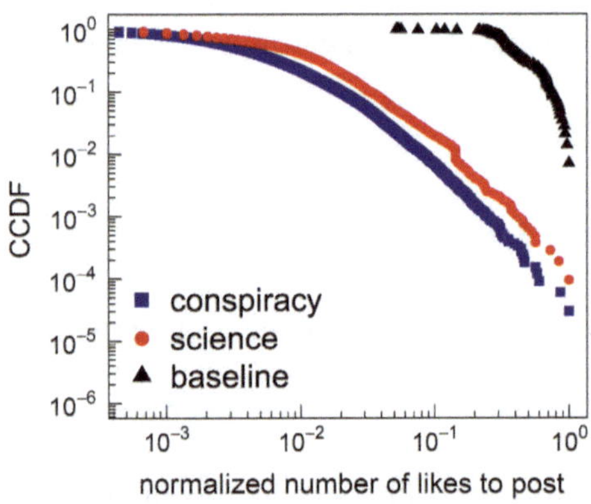

Fig. 3 Posts' consumption patterns. Complementary cumulative density function (CCDF) for the normalized number of likes received by posts belonging to science pages, conspiracy pages, and the baseline page. The CCDFs show remarkable differences between consumption patterns' distributions related to pages promoting heterogeneous contents and those related to the page promoting homogeneous contents

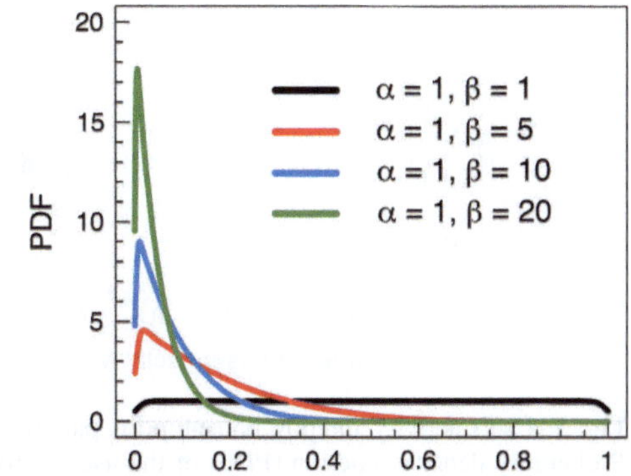

Fig. 4 Beta distribution $\mathscr{B}e(\alpha, \beta)$. Two parameters, α and β, control the shape of the distribution. In particular, for $\alpha = 1$ and $\beta = 1$ the Beta distribution $\mathscr{B}e(\alpha, \beta)$ is equivalent to the Uniform distribution $\mathscr{U}(0, 1)$. Conversely, if $\alpha = 1$ and $\beta \gtrsim 20$, the Beta distribution $\mathscr{B}e(\alpha, \beta)$ is a right heavy–tailed distribution

5.2 *Modeling Contents Consumption*

Here we introduce a model of pattern consumption that exploits the Beta distribution properties to generate different levels of posts' attractiveness, thus varying content–heterogeneity in the simulated collection of posts.

The Beta distribution is a family of continuous probability distributions defined in the interval [0, 1] and characterized by two real parameters, $\alpha > 0$ and $\beta > 0$, which control the shape of the distribution. In particular, for $\alpha = 1$ and $\beta = 1$ the Beta distribution $\mathscr{B}e(\alpha, \beta)$ is equivalent to the Uniform distribution $\mathscr{U}(0, 1)$. Conversely, if $\alpha = 1$ and $\beta \gtrsim 20$, the Beta distribution $\mathscr{B}e(\alpha, \beta)$ is a right heavy–tailed distribution. Figure 4 shows the Beta probability density function with respect to the two shape parameters α and β.

In our model, each post has a value drawn from a Beta distribution $v \sim \mathscr{B}e\,(1, \beta)$, with β ranging between 1 and 1,000,000, indicating its attractiveness. We let the parameter β assume those extreme values in order to obtain different distributions for posts' attractiveness. Indeed, notice that when $\beta = 1$ the Beta distribution $\mathscr{B}e\,(1, \beta)$ is equivalent to a uniform distribution $\mathscr{U}\,(0, 1)$, so that we have a collection of homogeneous–content posts—i.e., each post has the same degree of attractiveness; whereas when $\beta \to \infty$ the Beta distribution $\mathscr{B}e\,(1, \beta)$ is equivalent to a right heavy–tailed distribution, so that we have a collection of heterogeneous–content posts—i.e., there are few posts with a high level of attractiveness, while the vast majority of the posts is characterized by a low level of attractiveness. Moreover, each user is characterized by two parameters randomly drawn from power law distributions: her volume of activity, $a \sim p(x)$; and her fixed–preference about the posts, $b \sim p\,(x)$, where $p\,(x) = x^{-\gamma}$ with $\gamma = 1.5$. Each user can not exceed her assigned volume of activity, a, and she likes a given post if and only if her normalized[2] fixed–preference, b, is smaller than the attractiveness, v, of that post. Note that in our model we do not take into account the users' network: since Facebook network is very

[2]Note that we performed a unity–based normalization in order to bring all values of $b \sim p\,(x) = x^{-1.5}$ in the range [0, 1], so that the fixed–preference of the user is comparable with the attractiveness of the posts.

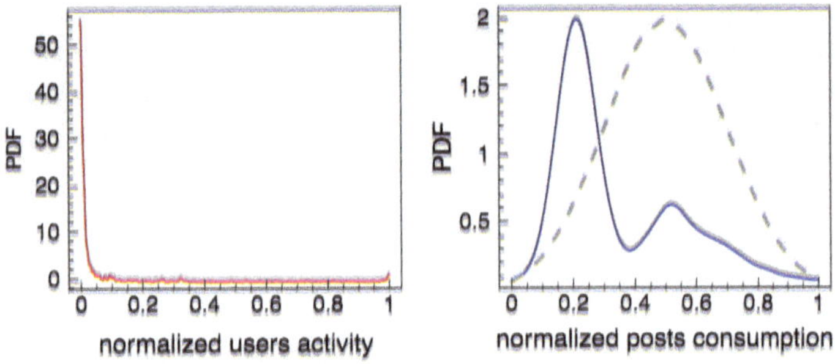

Fig. 5 Users activity and post consumption patterns with extremely heterogeneous–content posts. Probability density function (PDF) of the users activity and the posts consumption patterns generated by a simulation of the model with $\beta = 1{,}000{,}000$. If the content promoted by a page is heterogeneous, the heavy–tailed users' activity resolves in skewed posts consumption's patterns

dense—indeed, the diameter of Facebook social network is just 3.74 [5, 6]—the connections between users are not likely to influence posts' consumption dynamics.

We run simulations for β ranging between 1 and 1,000,000, with $P = 10{,}000$ (posts) and $U = 20{,}000$ (users). Results are averaged over 100 iterations.

Figure 5 shows the probability density function (PDF) of the users activity and the posts consumption patterns generated by a simulation of the model with $\beta = 1{,}000{,}000$—i.e., in the case of extremely heterogeneous–content posts. Observe that users' activity is heavy–tailed, and the distribution of posts' consumption is skewed. Such a result is consistent with empirical data shown in the previous section: if the content promoted by a page is heterogeneous, the heavy–tailed users' activity resolves in skewed posts consumption's patterns.

Figure 6 shows the probability density function (PDF) of the users activity and the posts consumption patterns generated by a simulation of the model with $\beta = 1$—i.e., in the case of homogeneous–content posts. Notice that users' activity is heavy–tailed,

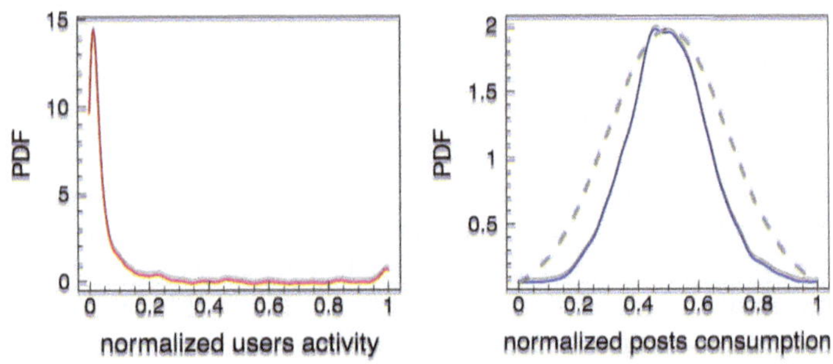

Fig. 6 Users activity and post consumption patterns with homogeneous–content posts. Probability density function (PDF) of the users activity and the posts consumption patterns generated by a simulation of the model with $\beta = 1$. If the content promoted by a page is always the same, the heavy–tailed users' activity resolves in approximately Gaussian posts consumption's patterns

whereas posts' consumption is approximately Gaussian. Such a result is consistent with empirical data shown in the previous section: if the content promoted by a page is always the same, the heavy–tailed users' activity resolves in approximately Gaussian posts consumption's patterns.

6 Concluding Remarks

Facebook is full by different and heterogeneous contents, ranging from the latest news all the way to satirical and funny stories. Each piece of content posted reflects the heterogeneity of the underlying social background of the over 1 Billion Facebook users. Online social networks such as Facebook and Twitter give people an outlet within which to express their attitudes, passions, and emotions by producing, sharing and, consuming heterogeneous information.

In Facebook, we found a fascinating case of contents' homogeneity: a page with more than 40 K followers that every day posts the same picture of Toto Cutugno, a popular Italian singer. In this work, we use such a page as a benchmark to investigate and model the effect that intrinsic contents heterogeneity has on popularity. In particular, we use that page for a comparative analysis of information consumption patterns with respect to pages posting heterogeneous contents related to Science and Conspiracy Theories, two topics with widely different audiences.

Surprisingly, we find that variations in the popularity of individual posts are due mostly to content heterogeneity. Even though there are no remarkable differences in user activity patterns between the Science, Conspiracy and Baseline pages, we observe that post popularity in the baseline page is well approximated by a normal distribution while it is broad tailed in pages promoting heterogeneous content. Finally, we show that these differences can be explained just by content heterogeneity by deriving a conceptually simple model that is able to reproduce our empirical observations.

Acknowledgements Funding for this work was provided by EU FET project MULTIPLEX nr. 317532 and SIMPOL nr. 610704. The funders had no role in study design, data collection and analysis, decision to publish, or preparation of the manuscript. We want to thank Prof. Guido Caldarelli for precious insights and contribution on the data analysis. Special thanks to Josif Stalin, Stefano Alpi, Michele Degani for giving access to the Facebook page of *La stessa foto di Toto Cutugno ogni giorno*. Bruno Gonçalves thanks the Moore and Sloan Foundations for support as part of the Moore-Sloan Data Science Environment at NYU.

References

1. Ebola e gattini. https://www.facebook.com/ebolagattini (2015). Accessed Jan 2015, facebook Page
2. A favore delle scie chimiche. https://www.facebook.com/afavoredellesciechimiche (2015). Accessed Jan 2015, facebook Page

3. La stessa foto di toto cutugno ogni giorno. https://www.facebook.com/totocutugno666 (2015). Accessed Jan 2015, facebook Page
4. Adamic, L., Glance, N.: The Political Blogosphere and the 2004 U.S. Election: Divided They Blog. In: LinkKDD '05: Proceedings of the 3rd International Workshop on Link Discovery, pp. 36–43 (2005)
5. Anagnastopolos, A., Bessi, A., Caldarelli, G., Del Vicario, M., Petroni, F., Scala, A., Zollo, F., Quattrociocchi, W.: Viral misinformation: the role of homophily and polarization. arXiv
6. Backstrom, L., Boldi, P., Rosa, M., Ugander, J., Vigna, S.: Four degrees of separation. In: Proceedings of the 4th Annual ACM Web Science Conference, pp. 33–42. WebSci '12, ACM, New York, NY, USA (2012). doi:10.1145/2380718.2380723
7. Bandari, R., Asur, S., Huberman, B.A.: The pulse of news in social media: forecasting popularity. In: ICWSM (2012)
8. Bauckhage, C.: Insights into internet memes. In: Proceedings of the Fifth International AAAI Conference on Weblogs and Social Media, pp. 42–49. AAAI (2011)
9. Ben-Naim, E., Krapivsky, P.L., Vazquez, F., Redner, S.: Unity and discord in opinion dynamics. Physica A (2003)
10. Bessi, A., Caldarelli, G., Del Vicario, M., Scala, A., Quattrociocchi, W.: Social determinants of content selection in the age of (Mis)information. In: Aiello, L., McFarland, D. (eds.) Social Informatics, Lecture Notes in Computer Science, vol. 8851, pp. 259–268. Springer International Publishing (2014). doi:10.1007/978-3-319-13734-6_18
11. Bessi, A., Coletto, M., Davidescu, G.A., Scala, A., Quattrociocchi, W.: Science vs Conspiracy: collective narratives in the age of misinformation. PLoS One (to appear)
12. Bessi, A., Scala, A., Zhang, Q., Rossi, L., Quattrociocchi, W.: The economy of attention in the age of (mis)information. J. Trust Manag. (to appear)
13. Bond, R.M., Fariss, C.J., Jones, J.J., Kramer, A.D.I., Marlow, C., Settle, J.E., Fowler, J.H.: A 61-million-person experiment in social influence and political mobilization. Nature **489**(7415), 295–298, Sept 2012. doi:10.1038/nature11421
14. Castellano, C., Fortunato, S., Loreto, V.: Statistical physics of social dynamics. Rev. Mod. Phys. **81**(2), 591, June 2009. doi:10.1103/RevModPhys.81.591
15. Centola, D.: The spread of behavior in an online social network experiment. Science **329**(5996), 1194–1197, Sept 2010. doi:10.1126/science.1185231
16. Cheng, J., Adamic, L., Dow, A.P., Kleinberg, J.M., Leskovec, J.: Can cascades be predicted? In: Proceedings of the 23rd International Conference on World Wide Web, pp. 925–936. WWW '14, International World Wide Web Conferences Steering Committee, Republic and Canton of Geneva, Switzerland (2014). doi:10.1145/2566486.2567997
17. Dawkins, R.: The Selfish Gene. Oxford University Press (1989)
18. Dukas, R., Kamil, A.C.: Limited attention: the constraint underlying search image. Behav. Ecol. **12**(2), 192–199 (2001)
19. Figueiredo, F., Almeida, J.M., Benevenuto, F., Gummadi, K.P.: Does content determine information popularity in social media?: a case study of youtube videos' content and their popularity. In: Proceedings of the 32nd annual ACM conference on Human factors in computing systems, pp. 979–982. ACM (2014)
20. Friggeri, A., Adamic, L., Eckles, D., Cheng, J.: Rumor Cascades. AAAI Conference on Weblogs and Social Media (ICWSM) (2013)
21. Goldhaber, M.H.: The attention economy and the net. First Monday **2**(4) (1997)
22. Gómez, V., Kaltenbrunner, A., López, V.: Statistical analysis of the social network and discussion threads in slashdot. In: Proceedings of the 17th international conference on World Wide Web, pp. 645–654. ACM (2008)
23. Gonzalez-Bailon, S., Borge-Holthoefer, J., Rivero, A., Moreno, Y.: The dynamics of protest recruitment through an online network. Sci. Rep. (2011)
24. Hannak, A., Margolin, D., Keegan, B., Weber, I.: Get back! you don't know me like that: the social mediation of fact checking interventions in twitter conversations. In: Proceedings of the 8th International AAAI Conference on Weblogs and Social Media (ICWSM'14). Ann Arbor, MI, June 2014

25. Kleinberg, J.: Analysis of large-scale social and information networks. Philos. Trans. R. Soc. A: Math. Phys. Eng. Sci. **371**, (2013)
26. Lehmann, J., Gonçalves, B., Ramasco, J.J., Cattuto, C.: Dynamical classes of collective attention in twitter. In: Proceedings of the 21st International Conference on World Wide Web, pp. 251–260. WWW '12, ACM, New York, NY, USA (2012). doi:10.1145/2187836.2187871
27. Lerman, K., Ghosh, R.: Information contagion: an empirical study of the spread of news on digg and twitter social networks. ICWSM **10**, 90–97 (2010)
28. Lerman, K., Hogg, T.: Using a model of social dynamics to predict popularity of news. In: Proceedings of the 19th international conference on World wide web, pp. 621–630. ACM (2010)
29. Leskovec, J., Backstrom, L., Kleinberg, J.: Meme-tracking and the dynamics of the news cycle. In: Proceedings of the 15th ACM SIGKDD International Conference on Knowledge Discovery and Data Mining, pp. 497–506. ACM (2009)
30. Lewis, K., Gonzalez, M., Kaufman, J.: Social selection and peer influence in an online social network. Proc. Nat. Acad. Sci. **109**(1), 68–72, Jan 2012. doi:10.1073/pnas.1109739109
31. Mocanu, D., Baronchelli, A., Gonçalves, B., Perra, N., Zhang, Q., Vespignani, A.: The twitter of babel: mapping world languages through microblogging platforms. PLoS One **8**(4), e61981. http://dblp.uni-trier.de/db/journals/corr/corr1212.html#abs-1212-5238 (2013)
32. Onnela, J.P., Reed-Tsochas, F.: Spontaneous emergence of social influence in online systems. Proce. Nat. Academ. Sci. **107**(43), 18375–18380, Oct 2010. doi:10.1073/pnas.0914572107
33. Paolucci, M., Eymann, T., Jager, W., Sabater-Mir, J., Conte, R., Marmo, S., Picascia, S., Quattrociocchi, W., Balke, T., Koenig, S., Broekhuizen, T., Trampe, D., Tuk, M., Brito, I., Pinyol, I., Villatoro, D.: Social Knowledge for e-Governance: Theory and Technology of Reputation. ISTC-CNR, Roma (2009)
34. Quattrociocchi, W., Caldarelli, G., Scala, A.: Opinion dynamics on interacting networks: media competition and social influence. Sci. Rep. **4**, May 2014. doi:10.1038/srep04938
35. Quattrociocchi, W., Conte, R., Lodi, E.: Opinions manipulation: media, power and gossip. Adv. Complex Syst. **14**(4), 567–586 (2011)
36. Quattrociocchi, W., Paolucci, M., Conte, R.: On the effects of informational cheating on social evaluations: image and reputation through gossip. IJKL **5**(5/6), 457–471. http://dblp.uni-trier.de/db/journals/ijkl/ijkl5.html#QuattrociocchiPC09 (2009)
37. Ratkiewicz, J., Conover, M., Meiss, M., Gonçcalves, B., Flammini, A., Menczer, F.: Detecting and tracking political abuse in social media. In: Proceedings of the Fifth International AAAI Conference on Weblogs and Social Media (2012)
38. Ratkiewicz, J., Conover, M., Meiss, M., Gonçalves, B., Flammini, A., Menczer, F.: Detecting and tracking political abuse in social media. In: ICWSM (2011)
39. Rojecki, A., Meraz, S.: Rumors and factitious informational blends: the role of the web in speculative politics. New Media Soc. http://nms.sagepub.com/content/early/2014/05/16/1461444814535724 (2014). Accessed May 2014
40. Salganik, M.J., Dodds, P.S., Watts, D.J.: Experimental study of inequality and unpredictability in an artificial cultural market. Science **311**(5762), 854–856 (2006)
41. Simon, H.: Designing organizations for an information-rich world. In: Computers, Communication, and the Public Interest, pp. 37–52 (1971)
42. Szabo, G., Huberman, B.A.: Predicting the popularity of online content. Commun. ACM **53**(8), 80–88 (2010)
43. Tapscott, D., Williams, A.D.: Wikinomics: How Mass Collaboration Changes Everything. Portfolio Hardcover (2006)
44. Tatar, A., de Amorim, M.D., Fdida, S., Antoniadis, P.: A survey on predicting the popularity of web content. J. Internet Serv. Appl. **5**(1), 1–20 (2014)
45. Tatar, A., Leguay, J., Antoniadis, P., Limbourg, A., de Amorim, M.D., Fdida, S.: Predicting the popularity of online articles based on user comments. In: Proceedings of the International Conference on Web Intelligence, Mining and Semantics, p. 67. ACM (2011)
46. Ugander, J., Backstrom, L., Marlow, C., Kleinberg, J.: Structural diversity in social contagion. Proc. Nat. Academ. Sci. http://www.pnas.org/content/early/2012/03/27/1116502109.abstract (2012)

47. Watts, D.J.: A simple model of global cascades on random networks. Proc. Nat. Academ. Sci. **99**(9), 5766–5771 (2002)
48. Weng, L., Flammini, A., Vespignani, A., Menczer, F.: Competition among memes in a world with limited attention. Sci. Rep. (2012)
49. Zadeh, A.H., Sharda, R.: Modeling brand post popularity dynamics in online social networks. Decis. Support Syst. (2014)

Part VII
Biological Networks

Investigating Side Effect Modules in the Interactome and Their Use in Drug Adverse Effect Discovery

Emre Guney

Abstract One of the biggest challenges in drug development is increasing costs of bringing new drugs to the market. Many candidate drugs fail during phase II and III trials due to unexpected side effects and experimental methods remain cost ineffective for large scale discovery of adverse effects. Alternatively, computational methods are used to characterize drug side effects, but they often rely on training predictors based on drug and side effect similarity. Moreover, these methods are typically tailored to the underlying data set and provide little mechanistic insights on the predicted associations. In this study, we investigate the role of network topology in explaining observed side effects of drugs. We show that the interactome based proximity can be used to identify side effects and we highlight a use case in which interactome-based side effect prediction can give insights on drug side effects observed in the clinic.

1 Introduction

Drug safety is one of the major driving factors beneath the attrition of drugs, contributing to more than 20% of the clinical trial failures and thus increasing costs associated with drug development [1, 2]. Undesired side effects of drugs are also among the leading causes of mortality in Western countries [3], prompting a clear need for better understanding of drug side effects.

The topology of the human interactome encodes biologically relevant information that can be used to discover novel drug-disease [4–6], and drug-side effect [7, 8] relationships. Although, some side effects can be explained by the proteins the drug is intended to target, many side effects likely to originate from the interactions of the drug with off-targets or the interactions between these proteins [9]. To understand the role of protein interactions in drug induced arrhythmias, Berger and colleagues identified the neighborhood of disease associated genes for long-QT syndrome in the PPI network and used this neighborhood to predict drugs likely to have risks for

E. Guney (✉)
Joint IRB-BSC-CRG Program in Computational Biology, Institute for Research in Biomedicine (IRB), c/ Baldiri Reixac 10-12, 08028 Barcelona, Spain
e-mail: emre.guney@irbbarcelona.org

B. Gonçalves et al. (eds.), *Complex Networks VIII*,
Springer Proceedings in Complexity, DOI 10.1007/978-3-319-54241-6_21

QT-interval prolongation [7]. They calculated a random-walk based score from each protein in the PPI network to known disease genes involved in long-QT syndrome, corresponding to the reachability of the proteins from the known disease genes. They then used this score to define a long-QT syndrome specific interactome neighborhood and to rank the drugs based on the targets falling in this neighborhood. Moreover, Brouwers et al., investigated whether the side effect similarity between drugs could be explained by the closeness of the drug targets in a functional PPI network [8]. They observed that only a minor fraction (6%) of drugs whose targets were direct neighbors in the network shared similar side effects, emphasizing the need for taking the global topology of the network into account.

In this study, we aim to investigate whether the global topology of the human interactome can characterize drug side effects. We first define side effect modules as the drug targets elucidating the side effects and check the network-based distances between side effect modules and drug targets. We show that drug targets are closer to the proteins associated with the known side effects of the drug in the network compared to the proteins associated with the rest of the side effects. We then use interactome based closeness to systematically identify side effects of the Federal Drug Administration (FDA) approved drugs in the DrugBank database. Finally, we demonstrate how the interactome based closeness can be used to predict side effects of tamoxifen that are not listed in SIDER.

2 Materials and Methods

2.1 Data Sets

The drugs used in our analysis were retrieved from DrugBank v4.3 database [10]. For all FDA approved drugs, we extracted drug-protein interactions including drug target, enzyme, transporter and carrier interactions (hereafter we simply refer all these proteins as drug targets). Uniprot ids from DrugBank were mapped to ENTREZ gene ids using Uniprot id mapping file (retrieved on October 2015). The SMILES strings of drugs were also downloaded from DrugBank.

We obtained drug side effect information from SIDER v4 [11], a resource containing side effects extracted from drug labels via text mining and mapped the drug ids to DrugBank ids using the PubChem mapping provided in DrugBank. We represented the side effects with their preferred terms reported in SIDER. To avoid including drugs whose side effects are not well characterized, we only considered drugs with at least five side effects in SIDER.

For validation purposes, in addition to SIDER, we used OFFSIDES [12], cataloging clinically significant drug side effects from FDA adverse event reporting system. We parsed the OFFSIDES flat file and mapped the drug ids to DrugBank ids using the PubChem mapping provided in DrugBank as we did for SIDER. Only the side effects with observed medical effect were included in the analysis.

We used the human interactome curated in a recent study [13], containing physical interactions between proteins from various large scale resources. The coverage and confidence of this integrated interaction network has been showed to be superior to interaction networks coming from yeast-two-hybrid or functional association data sets [6, 13]. Following the methodology in these studies, the largest connected component of the network, containing 141,150 interactions between 13,329 proteins, was used in the analysis.

2.2 *Defining Side Effect Modules*

To identify drug targets that contribute to the side effects, we followed the procedure presented in Kuhn et al. [14]. For each side effect and drug target we counted the number of drugs with and without the side effect for which the drug target was a known target versus the number of drugs with and without the side effect for which the target was not a known target. We used Fisher's exact test to calculate the two sided P-value of the observed occurrence of the target with the side effect as follows: The P-values were then corrected for multiple hypothesis testing using Benjamini and Hochberg's method. We selected the targets that were below 20% false discovery rate to describe the side effect module. In our analysis, we considered the side effects modules that had at least five targets in the interactome. We note that although the proposed approach is applicable to side effects defined by any number of proteins, we use the side effects with at least five proteins to ensure that the side effects in the analysis can be fairly explained by a group of proteins. We provide the side effect module information and the Jupyter Notebook to replicate the analysis in this study at http://www.github.com/emreg00/proxide.

2.3 *Characterizing Closeness Between Drug Targets and Side Effect Modules*

Given a network $G(V, E)$, we defined the following topological measures to quantify the network based closeness between targets of a drug, T, and proteins in a side effect module, S.

1. *Shortest*: The average pairwise shortest path length between each drug target and side effect module protein.

$$d_{\text{Shortest}}(T, S) = \frac{1}{\|T\| * \|S\|} \sum_{s \in S} \sum_{t \in T} d(t, s)$$

where $d(t, s)$ is the shortest path length between nodes t (a drug target) and s (a side effect protein) in the network. To convert the average shortest path length

above to a side effect specific z-score for each drug, we normalized $d_{\text{Shortest}}(T, S)$ using the mean ($\mu_{d_{\text{Shortest}}(T,S)}$) and standard deviation ($\sigma_{d_{\text{Shortest}}(T,S)}$) of $d_{\text{Shortest}}(T_i, S)$ values calculated for all the drugs $\{T_1, T_2, \ldots, T_n\}$ in the data set.
We used Dijkstra's shortest path algorithm implemented in Python networkx package to calculate the pairwise shortest path length between pairs of proteins in the interactome.

2. *Closest*: The average shortest path length to the closest protein in the side effect module from the drug targets, given by

$$d_{\text{Closest}}(T, S) = \frac{1}{\|T\|} \sum_{t \in T} \min_{s \in S} d(t, s)$$

We normalized these values using the mean and standard deviation of the values calculated for all the drugs as it was done above.

3. *PageRank*: The average PageRank score of the drug targets when the proteins in the side effect module were used to weight the influence of the nodes in the network. We assigned higher priors to the proteins in the side effect module, 1, compared to the rest of the nodes that were assigned 0.01 and calculated the probability that a random walker in the network would end up in a certain node based on the following formula:

$$PR_{i+1}(u) = (1-d) * PR_0(u) + d \sum_{v \in Neighbors(u)} \frac{PR_i(v)}{degree(v)}$$

where u was the current node in consideration, v was a node connected to u, $PR_i(u)$ was the PageRank score at iteration i and d is *damping factor* that was set to 0.15. The algorithm was repeated till convergence. The drug-side effect closeness was then defined using the PageRank score of the targets T normalized using the mean and standard deviation of the PageRank scores of all nodes for the given side effect. We used PageRank with priors implementation in GUILD package [15].

4. *NetScore*: The average NetScore score of the drug targets when the proteins in the side effect module were used as the source of information passed among the nodes. NetScore scored all the nodes in the network by iteratively propagating the score of the proteins in the side effect module to the neighboring nodes through shortest paths [15]. Unlike conventional shortest path based algorithms, NetScore considered the alternative shortest paths in between two nodes, favoring the nodes that were connected with more paths. We used the implementation of NetScore in GUILD software package [15] and initialized the proteins with a score of 1 if they belong to the side effect module and 0.01, otherwise. We limited the number of repetitions the program used to 3 with 2 iterations in each of them. The drug-side effect closeness was then defined as the average NetScore score of the targets T normalized using the mean and standard deviation of the NetScore scores of all nodes for the given side effect.

5. *Proximity*: The significance of the observed average shortest path length to the closest protein in the side effect module from the drug targets. Interactome based proximity [6] first quantified the average shortest path length between the closest protein in the side effect module and the drug targets ($d_{\text{Closest}}(T, S)$ above) and then calculated a z-score corresponding significance of these distances using the mean and the standard deviation of the background distribution of expected minimum shortest path distances between two randomly selected groups of proteins (with the same size and degrees of the original protein sets). The background distance distribution was generated using 1,000 randomly selected protein groups matching drug targets and side effect proteins.

2.4 Drug Side Effect Prediction Using Network-Based Closeness

To investigate whether the network-based closeness can predict side effects, for each known and unknown drug and side effect pair, we recorded the five topology based closeness scores ($z_{Shortest}$, $z_{Closest}$, $z_{PageRank}$, $z_{NetScore}$, $z_{Proximity}$). We then verified whether these topology based scores could discriminate known drug-side effect pairs from the rest by calculating the number of correctly and incorrectly predicted known and unknown drug-side effect pairs at various score cutoffs and checking the area under ROC curve (AUROC) and area under precision-recall curve (AUPRC). The known drug-side effect associations in SIDER and OFFSIDES databases were used as the gold standard positive instances and the remaining associations were assumed to be negative instances. We employed Python scikit-learn package to calculate AUROC and AUPRC values and R for the statistical tests.

3 Results

3.1 Side Effect Modules in the Interactome

The available experimental information on the drug targets contributing to the side effects of drugs is often limited to a handful of drug targets [16, 17], hindering a large scale analysis of drug targets inducing the side effects. Alternatively, over-representation analysis of drug targets and side effects can characterize the targets eliciting side effects [14]. Therefore, we define the side effect modules as the groups of drug targets significantly associated with the side effects using the drug target information in DrugBank [10] and SIDER database [11]. Using 1,530 FDA approved drugs and their targets in DrugBank, we identify 1,177 drug target groups associated with the side effects. To confirm that the proteins defining the side effect modules are biologically relevant, we check the overlap between the side effect targets by Lounkine et al. [17]. The side effect modules cover at least one protein associated

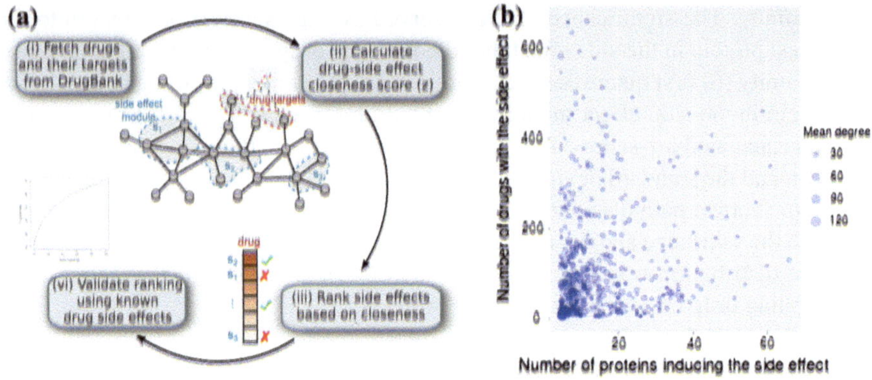

Fig. 1 Side effect modules in the interactome and their use in drug adverse effect characterization. **a** Schematic overview of the interactome based analysis of drug side effect modules. For each of 817 drugs and 537 side effects, we calculate network based closeness between the drug targets and the proteins inducing the side effect and validate the predictions using known drug-side effect associations. **b** Each point represents a side effect consisting of proteins identified to be significantly associated to the side effect. The x-axis is the number of proteins in the side effect module and the y-axis is the number of drugs that shows the side effect. The size of the points scales with the median degree of the proteins in the side effect module

with the side effect for 164 of 241 side effects that are also in the Lounkine et al. study. Furthermore, 130 out 265 of the proteins in the identified side effect modules appear among 224 proteins given in the Lounkine data set, covering more than half of the experimentally verified side effect targets.

To understand the interactome based relationship between drug targets and side effect modules, we focus on 537 side effect modules that have at least 5 proteins in the interactome and 817 drugs both known to exert any of these side effects and having at least one target in the interactome. We seek whether topological characteristics of these two groups of nodes, drug targets and side effect module proteins, can explain observed side effects of drugs (Fig. 1a). We first turn our attention to the side effect module proteins and ask if the number of proteins in the module or their degree can provide insights on the side effects drugs show. The average module size is $\langle n_{module} \rangle = 15.8$ among 537 side effects and the largest module, the one of gynaecomastia (enlargement of a man's breasts), contains 66 proteins. Interestingly, the average degree of all the proteins contributing to a side effect is higher than the average degree of the remaining proteins in the interactome ($\langle k_{sideeffect} \rangle = 26.5$ vs. $\langle k_{nonsideeffect} \rangle = 21.1$). If the proteins within each side effect module are considered independently, however, the average degree of the proteins in the side effect modules is around the average degree of the interactome ($\langle k_{module} \rangle = 20.8$ vs. $\langle k \rangle = 21.2$), with peliosis hepatis, an uncommon vascular condition in liver, being the side effect with the highest average degree $\left(\langle k_{peliosishepatis} \rangle = 123.6\right)$.

To investigate whether the size and the average degree of the identified side effect modules are higher for the "popular" side effects—the side effects that occur frequently in SIDER, we look at the number of drugs the side effect is observed and the number and mean degree of the proteins in the side effect module (Fig. 1b). The significant but low correlation between the number of drugs showing the side effects and the module size (Spearman's rank correlation $\rho = 0.16$, $P = 1.8 \times 10^{-4}$) suggests that the size of the module is not strongly associated to the occurrence of the side effects. On the other hand, the degree of the proteins within the side effect modules is not correlated with the number of drugs the side effect is observed (Spearman's rank correlation $\rho = 0.03$, $P = 0.55$).

3.2 Network Based Closeness of Drugs and Side Effects

Next, for each drug and side effect pair in our analysis (817 × 537 pairs), we calculate the network based closeness of the drug's targets to the side effect module in the interactome using five topological measures (see Methods). We then investigate how well the calculated closeness scores discriminate the observed drug side effects using the known drug side effect associations in SIDER and OFFSIDES databases (Table 1).

We find that the drugs tend to be closer to the proteins inducing the side effects known to be associated with them compared to the proteins in the rest of the side effect modules (Fig. 2). The difference in the closeness values of known and unknown drug-side effect pairs is significant using both SIDER and OFFSIDES side effect associations (one-sided MannWhitney U test $P \ll 0.05$). We observe that NetScore, the method that takes alternative shortest path between drug targets and side effect module proteins and Proximity, the method that compares observed shortest path length between drug targets and the closest side effect module protein to the distances between randomly selected nodes in the network yield a wider range of closeness scores than the remaining methods.

We then turn to predicting drug side effects using the network neighborhood information of the side effect modules and quantify the closeness between drug targets and side effect modules in the interactome. We use the drug-side effect associations in SIDER and OFFSIDES as the gold standard data to calculate the precision, recall, false positive rate at various closeness score cutoffs and check the area under the

Table 1 Number of drugs, side effects and known drug-side effect associations included in the analysis according to SIDER and OFFSIDES databases

	SIDER	OFFSIDES
Number of drugs	817	269
Number of side effects	537	118
Number of known drug-side effect associations	64,885	2,060
Percentage of known associations	14.8%	6.5%

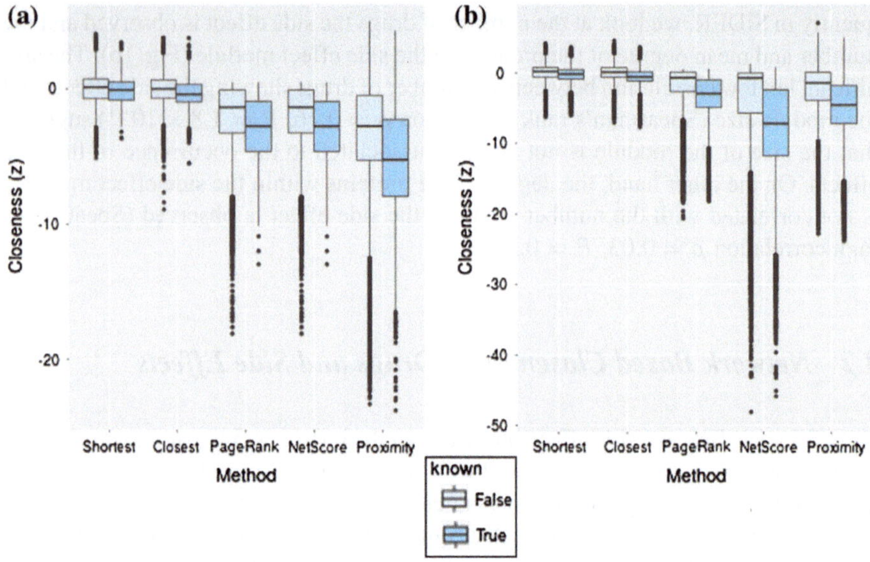

Fig. 2 Network based closeness of known and unknown drug-side effect pairs. The closeness between drug targets and side effects calculated using five topological measures (Closest, Shortest, PageRank, NetScore and Proximity) for each of 817 drugs and 537 side effects. Known drug-side effect associations are taken from **a** SIDER and **b** OFFSIDES

Table 2 AUROC, AUPRC and percentage of correctly predicted highest ranked drug-side effect pair for various network based closeness methods using SIDER and OFFSIDES associations

	AUROC (%)		AUPRC (%)		Correct at top (%)	
	SIDER	OFFSIDES	SIDER	OFFSIDES	SIDER	OFFSIDES
Shortest	59.8	53.9	17.8	7.1	15.9	8.2
Closest	67.9	57.7	27.6	8.5	79.6	28.6
PageRank	69.0	59.6	27.0	8.6	55.8	13.0
NetScore	71.7	61.9	28.8	9.6	52.1	14.5
Proximity	71.1	63.6	32.8	11.4	56.7	11.5

ROC curve (AUROC), the area under the precision-recall curve (AUPRC) and the percentage of the drugs for which the highest scoring prediction is a known side effect (Table 2). We see that, overall, the best performing methods are NetScore and Proximity, showing higher prediction accuracy on both SIDER and OFFSIDES data sets compared to the rest of the methods.

Despite using only the network topology the AUROCs for NetScore and Proximity scores on SIDER drug-side effect associations are 71.7% and 71.1%, respectively, suggesting that closeness of drugs to side effect modules is predictive of the drug's adverse effects. We also examine the area under precision-recall curve (AUPRC) and find that NetScore and Proximity achieve AUPRC values of 28.8% and 32.8%, respectively. Furthermore, for 52.1% and 56.7% of the drugs used in the analysis, the highest scoring side effect identified by NetScore and Proximity is reported in SIDER, showing that drug-side effect module closeness can provide insights on the side effects of drugs. On the other hand, when the drug-side effect associations in OFFSIDES database is used, the AUROC drops to 61.9% and 63.6% for NetScore and Proximity, still substantially higher than that would be expected from a classifier producing random predictions (50%). Moreover, only for around 10% of the drugs, the highest scoring side effect is in OFFSIDES, an observation we attribute to the lower coverage of known side effects in OFFSIDES database (6.5%) compared to the SIDER (14.8%, Table 1). Accordingly, due to the higher coverage of drugs and side effects, and better prediction accuracy, in the rest of the text, we use SIDER drug-side effect associations as the gold standard.

3.3 Assessing the Effect of the Data Incompleteness

The current knowledge on drug-target interactions represent only a partial view of the possibly many proteins involved in drug's action [18]. To account for the potential implications of incompleteness of the drug target data, we analyze the prediction performance of each method on various subsets of drugs and side-effects categorized with respect to the number of drug targets (m) and side effect proteins (n). Figure 3 shows the AUROC and AUPRC values *(i)* on the original data set containing 817 drugs with at least one target and 537 side effect modules of at least five proteins ($m \geq 1, n \geq 5$) and when we repeat the analysis using *(ii)* 428 drugs and 537 side effects with at least five targets and proteins ($m \geq 5, n \geq 5$), *(iii)* 428 drugs with at least five targets and 322 side effect modules with at least ten proteins ($m \geq 5$, $n \geq 10$), and finally, *(iv)* 176 drugs and 322 side effects with at least ten proteins ($m \geq 10, n \geq 10$).

We find that, as the drugs and side effects associated with more proteins are used, the closeness based predictions improve. Nonetheless, the improvement mainly stems from the higher number of drug targets, as the change in the accuracy is modest when the number of proteins in the side effect modules increases. On the other hand, the AUROC and AUPRC values increase 3–6% when the drugs with more number of targets are used.

3.4 Case Study: Top Ranking Side Effects of Tamoxifen

To highlight how interactome based closeness of drug targets can help identifying side effects, we use Proximity, the method that show high overall accuracy according to various performance measures (Table 2). Using only the target information of a given drug, Proximity calculates a network topology based significance of the closeness of the drug to all side effects, allowing us to rank the likelihood of all side effects for any drug with drug target information. Notably, among the drugs in our data set

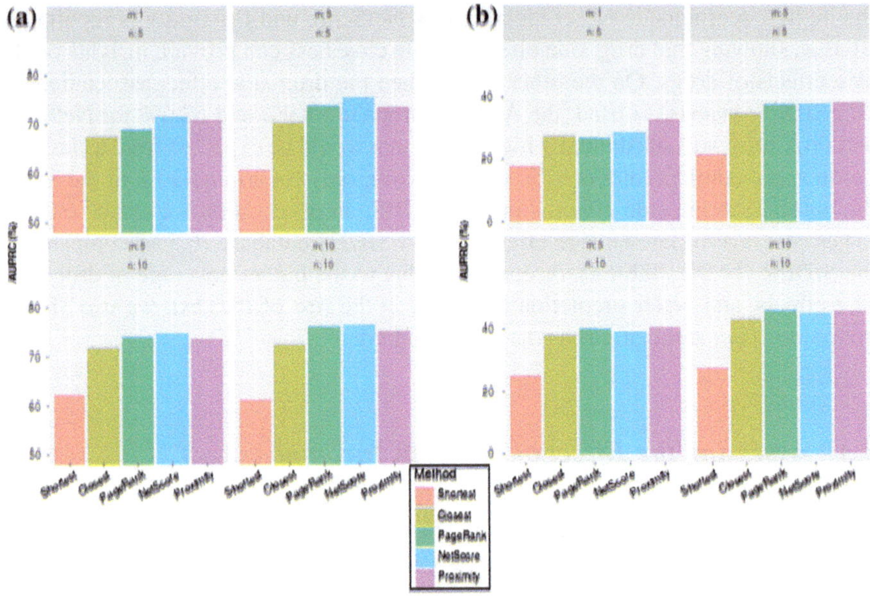

Fig. 3 The effect of data incompleteness on prediction performance. The area under **a** the ROC curve (AUROC) and **b** the precision-recall curve (AUPRC) values when a subset of the drugs and side effects are excluded from the analysis. In each panel, the drugs having less than *m* targets in the network and the side effect modules that have less than *n* proteins in the network are excluded from the analysis

for which the top ranking side effect is not reported in SIDER, we see tamoxifen, an estrogen receptor modulator used for the treatment of breast cancer. Although eight out of ten highest scoring side effects are reported in SIDER, two side effects with very strong association scores, "muscular weakness" and "neuropathy peripheral" are not listed in SIDER. We find out that the muscle weakness is indeed a known side effect according to the drug information in Medlineplus (http://www.nlm.nih.gov/medlineplus/druginfo/meds/a682414.html). Furthermore, while not indicated in neither SIDER nor Medlineplus, the peripheral neuropathy appears to be a clinically relevant condition reported by several patients in message boards (at http://www.community.breastcancer.org and http://www.medhelp.org).

The Proximity score of Tamoxifine to the 14 proteins associated to peripheral neuropathy is $z = -12.1$, suggesting that the drug targets are highly proximal to the side effect proteins in the interactome as a group. This is largely due to seven enzymes (*CYP1A2, CYP2C19, CYP2C8, CYP2C9, CYP2D6, CYP3A4, CYP3A7*) and two transporters (*ABCB1, ABCC2*) tamoxifen is known to bind are in the side effect module. Furthermore, protein encoded by *KIT* gene in the side effect module, is known to be inhibited via phosphorylation by Protein kinase C protein family, a family of proteins targeted by tamoxifen, contributing to the observed proximity to the peripheral neuropathy.

4 Conclusion

Most existing approaches rely on existing drug side effect associations to predict drug side effects, hindering both the interpretability of predicted associations and the ability to discover novel side effects. In contrast, in this study, we investigate the network based closeness of drug targets to the proteins likely to induce the side effects to explain the observed drug adverse effects. We use the interactome based closeness to predict side effects associated with a drug, providing a mechanistic explanation of the predicted association.

One drawback of network based methods is that they require that at least a drug target known to interact with a protein in the interactome. Furthermore, they can only be applied to side effects for which a set of proteins inducing the side effect can be identified. Yet, we show that interactome based closeness can systematically detect side effects of 817 FDA approved drugs in DrugBank without relying on the known drug-disease associations. Moreover, network based closeness offers an important advantage over widely used similarity based methods by providing interactome-based insights on the likelihood of a drug to induce a given side effect.

Acknowledgements The author is grateful to Dr. Patrick Aloy for providing computational resources for this study and the members of the lab for fruitful discussions. EG is supported by EU-cofunded Beatriu de Pinós incoming fellowship from the Agency for Management of University and Research Grants (AGAUR) of Government of Catalunya.

References

1. Allison, M.: Reinventing clinical trials. Nat. Biotechnol. **30**(1), 41–49 (2012)
2. Hay, M., Thomas, D.W., Craighead, J.L., Economides, C., Rosenthal, J.: Clinical development success rates for investigational drugs. Nat. Biotechnol. **32**(1), 40–51 (2014)
3. Tai-Yin, W., Jen, M.-H., Bottle, A., Molokhia, M., Aylin, P., Bell, D., Majeed, A.: Ten-year trends in hospital admissions for adverse drug reactions in England 1999–2009. J. R. Soc. Med. **103**(6), 239–250 (2010)
4. Zhao, S., Li, S.: A co-module approach for elucidating drug-disease associations and revealing their molecular basis. Bioinformatics **28**(7), 955–961 (2012)
5. Guney, E., Garcia-Garcia, J., Oliva, B.: GUILDify: a web server for phenotypic characterization of genes through biological data integration and network-based prioritization algorithms. Bioinformatics (Oxford, England) **30**(12), 1789–1790 (2014)
6. Guney, E., Menche, J., Vidal, M., Barabási, A.-L.: Network-based in silico drug efficacy screening. Nat. Commun. **7**, 10331 (2016)
7. Berger, S.I., Ma'ayan, A., Iyengar, R.: Systems pharmacology of arrhythmias. Sci. Signal. **3**(118), ra30 (2010)
8. Brouwers, L., Iskar, M., Zeller, G., van Noort, V., Bork, P.: Network neighbors of drug targets contribute to drug side-effect similarity. PLoS ONE **6**(7), e22187 (2011)
9. Berger, S.I., Iyengar, R.: Role of systems pharmacology in understanding drug adverse events. Wiley Interdiscip. Rev.: Syst. Biol. Med. **3**(2), 129–135 (2011)
10. Law, V., Knox, C., Djoumbou, Y., Jewison, T., Guo, A.C., Liu, Y., Maciejewski, A., Arndt, D., Wilson, M., Neveu, V., Tang, A., Gabriel, G., Ly, C., Adamjee, S., Dame, Z.T., Han, B., Zhou,

Y., Wishart, D.S.: DrugBank 4.0: shedding new light on drug metabolism. Nucleic Acids Res. **42**(Database issue), D1091–1097 (2014)
11. Kuhn, M., Letunic, I., Jensen, L.J., Bork, P.: The SIDER database of drugs and side effects. Nucleic Acids Res. **44**(D1), D1075–1079 (2016)
12. Tatonetti, N.P., Ye, P.P., Daneshjou, R., Altman, R.B.: Data-driven prediction of drug effects and interactions. Sci. Transl. Med. **4**(125), 125ra31 (2012)
13. Menche, J., Sharma, A., Kitsak, M., Ghiassian, S.D., Vidal, M., Loscalzo, J., Barabási, A.-L.: Disease networks. Uncovering disease-disease relationships through the incomplete interactome. Science (New York, N.Y.) **347**(6224), 1257601 (2015)
14. Kuhn, M., Al Banchaabouchi, M., Campillos, M., Jensen, L.J., Gross, C., Gavin, A.-C., Bork, P.: Systematic identification of proteins that elicit drug side effects. Mol. Syst. Biol. **9**(1), 663 (2013)
15. Guney, E., Oliva, B.: Exploiting Protein-protein interaction networks for genome-wide disease-gene prioritization. PLoS ONE **7**(9), e43557 (2012)
16. Ji, Z.L., Han, L.Y., Yap, C.W., Sun, L.Z., Chen, X., Chen, Y.Z.: Drug Adverse Reaction Target Database (DART): proteins related to adverse drug reactions. Drug Saf. **26**(10), 685–690 (2003)
17. Lounkine, E., Keiser, M.J., Whitebread, S., Mikhailov, D., Hamon, J., Jenkins, J.L., Lavan, P., Weber, E., Doak, A.K., Côté, S., Shoichet, B.K., Urban, L.: Large-scale prediction and testing of drug activity on side-effect targets. Nature **486**(7403), 361–367 (2012)
18. Mestres, J., Gregori-Puigjané, E., Valverde, S., Solé, R.V.: Data completenessthe Achilles heel of drug-target networks. Nat. Biotech. **26**(9), 983–984 (2008)

Attractor Analysis of the Asynchronous Boolean Model of the Klotho Gene Regulatory Network

Malvina Marku, Inva Koçiaj, Klotilda Nikaj and Margarita Ifti

Abstract When analyzing the state space attractors and their basin of attraction, two main methods have been used: the synchronous update method and the asynchronous update method. Although its simplicity, the synchronous update fails to consider various time scales of the chemical processes between the components within the network. To overcome this limitations, several asynchronous methods have been developed. In this work, we use two different asynchronous methods to study the dynamics of the wild-type and perturbed Klotho gene and carry out comparative results with previously published results. Prior evidence shows that the system develops oscillations and a fixed point. The numerical results of the both asynchronous methods show that, the system's oscillations disappear and the system undergo in only one time-invariant fixed point, leading to an extended attractor. This work aims to highlight different behavior of the Klotho gene under various internal and external perturbations.

Keywords Gene-regulatory network · Klotho gene · Boolean model · Asynchronous update · Fixed point

1 Introduction

In system biology, different formal presentation types are used to construct mathematical models, varying from quantitative continuous models to qualitative discrete models, each with their own strength and weakness. Continuous models (Ordinary Differential Equations ODE of the system) are often restricted by the absence or low availability of kinetic parameters and/or experimental data. On the other hand, discrete models, such as Boolean models, give the possibility to study large networks, but, still preventing some important properties of the system [1]. To bridge the gap

M. Marku (✉) · I. Koçiaj · K. Nikaj · M. Ifti
Faculty of Natural Sciences, University of Tirana, Tirana, Albania
e-mail: malvina.marku@fshn.edu.al

B. Gonçalves et al. (eds.), *Complex Networks VIII*,
Springer Proceedings in Complexity, DOI 10.1007/978-3-319-54241-6_22

between these two models, different models have been developed, such as hybrid models [2], in which every node is characterized by two variables: a continuous concentration and a discrete activity. Which model is to be selected to suit a certain system, depends on the level of quantitative details of the available experimental data [3, 4].

Boolean models were first introduced by Kauffman (1969) [5] and Thomas [6] for modelling gene-regulatory networks [3, 4, 7, 8] in a total absence of kinetic details. Since then, different applications provided adequate justifications for the use of Boolean models, since the input-output curves of regulatory interactions can be well-suited with step functions [9]. In this representation of a gene regulatory network, each node represents a gene, protein, chemical element, enzyme, etc., while each edge represents the chemical process between the nodes. The chemical processes consist only in the activation or inhibition form, while the activity of each node can take only two possible values: 1 (ON) (expressed, open gate, concentration above threshold, active) or 0 (OFF) (not expressed, closed gate, concentration below threshold, inactive). The future state of each node is determined by the current state of its regulators, through a Boolean function, usually expressed via the logic operators AND, OR and NOT.

Generating the future state of the each node, i.e., of the system, can undergo in two possible methods: the synchronous method or the asynchronous method. The simplest method, the synchronous updating method assumes equal time scales for all the processes involved and is implemented by updating all the nodes simultaneously [3, 4]. However, despite its simplicity, this method fails to consider different time scales of molecular interactions. The need to increase the accuracy of the method lead to the development of two different asynchronous methods, first introduced by Thomas [6]: the stochastic update method (Random Order Asynchronous Method and the General Order Asynchronous Method) and the deterministic update method (Deterministic Asynchronous Method). In a stochastic update method, in every time step, a random node/sequence of nodes is selected to update and the system is updated according to this node/sequence of nodes, while in a deterministic update method the system is updated according to a pre-determined sequence or at multiples of their pre-selected time units [2, 3].

In the majority of studies, the dynamic analysis of a Boolean network includes the analysis of the attractors (mainly interested on fixed points) in the state space of the system, their corresponding basin of attraction, starting from the wild-type condition. However, considering all the possible initial states and identifying cycles is still an intriguing study. In this work, we study the attractors of the Klotho gene networks by applying the Random Order Asynchronous (RA) and the General Order Asynchronous (GA) methods, and compare the results with the Synchronous method, obtained in previous studies [10].

Klotho, first discovered by Kuro-o in 1997, is the only reported single gene mutation that leads to a premature aging phenotype in mice [1, 11, 12]. From experimental data results that over-expression of Klotho leads to life span and low-expression of it leads to a numerous diseases, including Chronic Kidney Disease (CDK), hyperphosphatemia, hypercalcitrosis, vascular calcification, bone abnormalities, etc. [11–13]. Here, we focus on the role of Klotho in phosphate and calcium metabolism. Phosphate metabolism is regulated by several endocrine factors, involving vitamin D and parathyroid hormone (PTH). The active form of vitamin D is synthesized in kidney and acts on intestine to increase absorption of dietary calcium and phosphate. PTH acts on kidney to promote both vitamin D synthesis and phosphate excretion into urine. As a result, unlike vitamin D, PTH can selectively increase blood calcium levels without increase in blood phosphate levels.

Recent studies have identified fibroblast growth factor 23 (FGF23) as a novel endocrine factor that lowers blood phosphate and vitamin D levels. FGF23 is secreted from osteocytes in response to high blood level of phosphate and vitamin D. FGF23 acts on kidney to (1) induce phosphate excretion into urine, and (2) counter regulate vitamin D. As a result, FGF23 reduces phosphate absorption from intestine. Thus, FGF23 induces a negative phosphate feedback by functioning as a phosphaturic hormone, as well as a counter regulatory hormone of vitamin D.

Experiment results show that FGF23 and Klotho may function in a common signal transduction pathway. In fact, the transmembrane Klotho protein was shown to form a complex with several FGF receptors, e.g., Klotho functions as an obligatory co-receptor for FGFR, allowing FGF23 to suppress PTH. As a result, Klotho functions as both a phosphate regulatory hormone and a calcium regulatory hormone, through a cascade of interactions [11–13]. The simplified network of interactions between these components is given in Fig. 1 [14].

2 Methods

Let i, $(i = 1, 2, \ldots, N)$ represent the nodes of a biological network with N nodes. The basis of a Boolean model stands on the assumption that the state of each node i, at a certain time t, $X_i(t)$ can be determined by the states of its regulators at previous time, through a Boolean function. In order to implement time, first, we have to specify which the previous time is and what time is requested for any interaction to occur. In this direction, two different methods have been developed: the synchronous method [5] and the asynchronous methods [2, 3, 7, 8]. The synchronous method assumes similar time intervals for every interaction, i.e., all the nodes within the network are updated simultaneously:

$$X_i(t+1) = F_i(X_1(t), X_2(t), \ldots, X_n(t)) \quad (1)$$

where, F_i is the Boolean function of node i and $X_1, X_2, \ldots, X_n$ are the regulators of node i. Note that, a network of N nodes, can have a finite number of 2^N states.

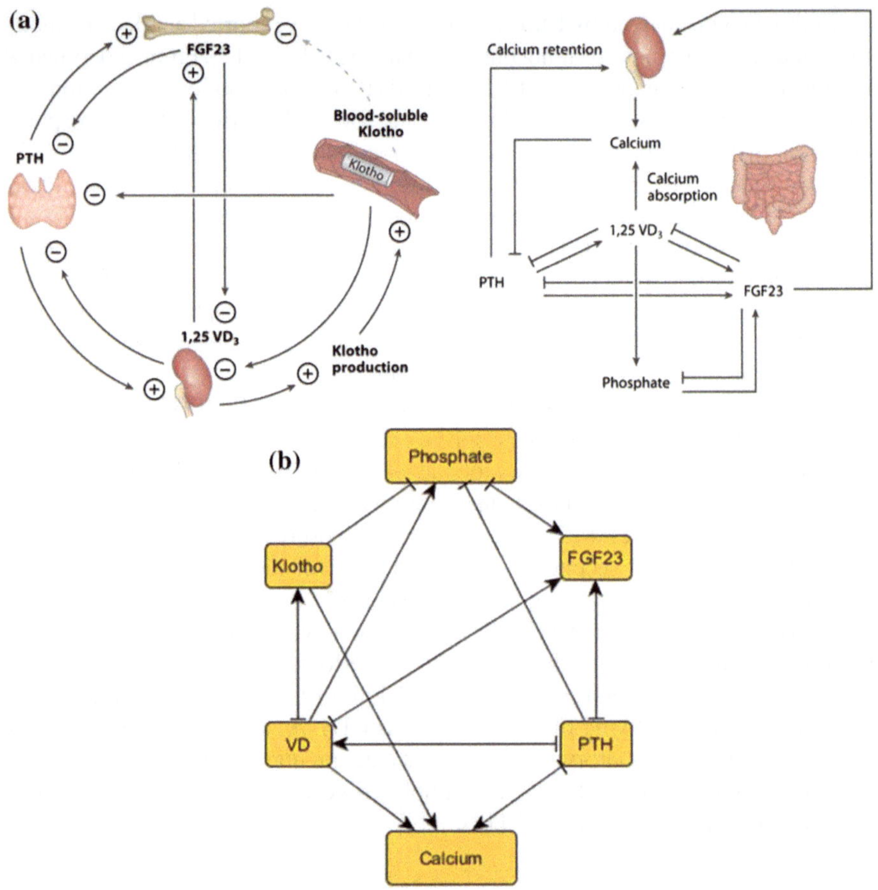

Fig. 1 The simplified aging network. **a** Elements interactions [2]. A change in the concentrations of one component, may lead to a cascade of events, starting with hypocalcemia. **b** Network representation: The nodes of this network represent proteins or ion channels, while the edges of the network (all directed) represent interactions between nodes (protein protein interaction, chemical reactions, and other indirect regulatory relationships between nodes). $\rightarrow$ denotes activation, while $\dashv$ denotes inhibition

It is clear to see that this method is very easy to be implemented in a certain network; even so, it does not take into account different time scales of different chemical processes (direct/indirect reactions, translational/transcriptional regulations, etc.), which scale from milliseconds to hours to occur. In order or add more information about the node characteristics, different asynchronous methods have been developed, wherein each node is updated up to their individual time scales.

1. ***Random Order Asynchronous (RA)***: at each time step, a random sequence of nodes from the normal distribution of the permutation space $\{1, 2, \ldots, N\}$ of $N!$ possible permutations of N nodes, is selected to update, and all the nodes are updated in that

order. In this case, the future state of node i can be determined according to the most recent state of its regulators during time interval $\tau_i \in [t, t+1]$:

$$X_i(t+1) = F_i\left(X_{i_1}\left(\tau_{i_1}\right), X_{i_2}\left(\tau_{i_2}\right), \ldots, X_{i_n}\left(\tau_{i_n}\right)\right) \tag{2}$$

If the position of regulator j is before node i, $\tau_{i_j} = t+1$, otherwise $\tau_{i_j} = t$.
2. ***General Order Asynchronous (GA)***: at each time step, a random node is selected to update.
3. ***Deterministic Asynchronous (DA)***: each node is characterized by a certain time unit γ_i and it can be updated only if the $t = k \cdot \gamma_i$:

$$X_i(t+1) = \begin{cases} F_i\left(X_1(t), X_2(t), \ldots, X_n(t)\right) \text{ if } t = k \cdot \gamma_i \\ X_i(t) \qquad\qquad\qquad\qquad \text{otherwise} \end{cases} \tag{3}$$

However the method used, dynamical analysis of a biological network is focused in studying the stability of the system, i.e., identifying the attractors of the system, which can be single fixed points or complex attractors, wherein the system oscillates among a certain number of states. The stability of the system can be studied in two common ways: (1) by solving the system of Boolean equations (note that, the fixed points are invariant to time, so we need to solve the system of equations after removing time), (2) from the state transition graph wherein each node represents a certain state and each edge represents its successor. The basin of attraction represents all the states leading to the attractor. Note that, in the synchronous updating method, each state has only one successor, therefore the basins of attraction of different attractors do not intersect, while in the asynchronous updating methods, the basins of attraction of different attractors overlap [4].

The Boolean system of equations associated with the Klotho network given in Fig. 1, is given in Table 1, where the asterisk denotes the future state of each node. Previous results obtained in [10] for the synchronous method show that the system can undergo in two possible attractors: a fixed point (010100) with a basin of attraction of 13% of nodes, and a limit cycle of length 4, with a basin of attraction 87% of the possible states. The reason why we are more interested in the fixed point is that, since it satisfies the equation $x^* = x$, it is time-invariant, and, therefore, it should remain un-affected by the updating method used.

Table 1 Boolean rules governing the nodes states in the 6-node network represented in Fig. 1b

Node	Boolean rule
Calcium	Calcium* = VD or Klotho and PTH
FGF23	FGF23* = PTH or VD or Phosphate
Klotho	Klotho* = VD
PTH	PTH* = not (VD or Calcium and FGF23)
Phosphate	Phosphate* = VD and not (Klotho or FGF23) and not PTH
VD	VD* = PTH and not (FGF23 or Klotho)

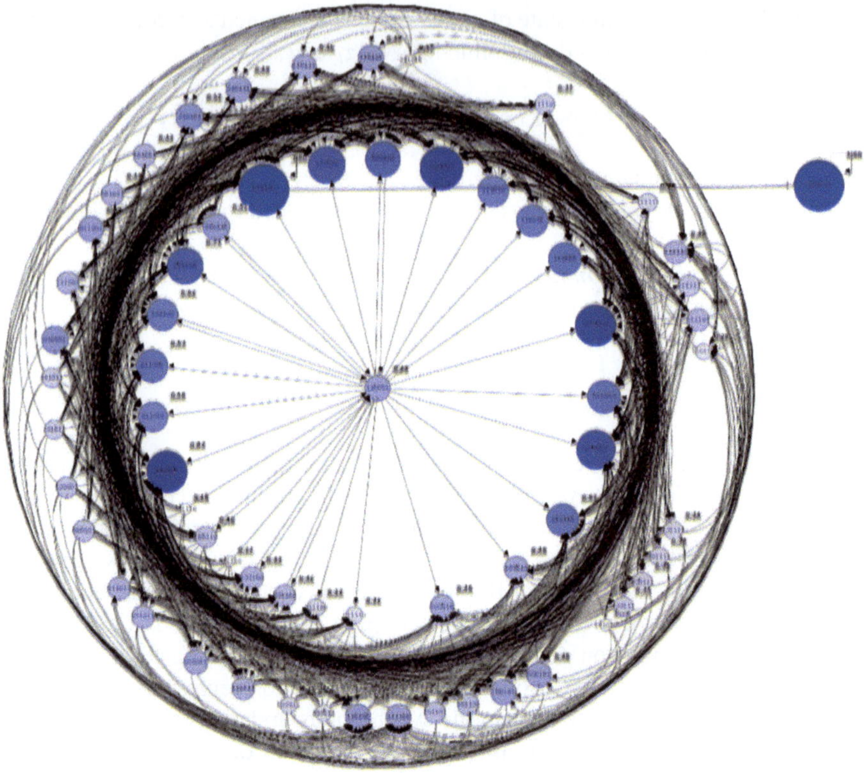

Fig. 2 The state transition graph obtained by applying RA method. The *digit numbers* represent the activity of each component, in the order: Calcium, FGF23, Klotho, PTH, Phosphate, VD. The fixed point (010100) is individualized on the *right*

In the next step, we applied the ROA and the GA methods in the case of hypocalcemia (Calcium = False). Because these methods are stochastic and are based in random selection of the nodes/sequences, the simulations are taken for a large number of time steps. The results are shown in Figs. 2 and 3. As we see, the fixed point of the system, in both methods, remains the same [11, 15].

3 Results and Discussions

As we see from our results, the dynamic analysis of the Klotho gene network gives us the same fixed point (010100), as the synchronous method did. Even that our model is a toy model (to obtain more precise results, we need to consider more interacting components and to detail the way the components interact with each—other through different processes; this study will be published in future works), the biological

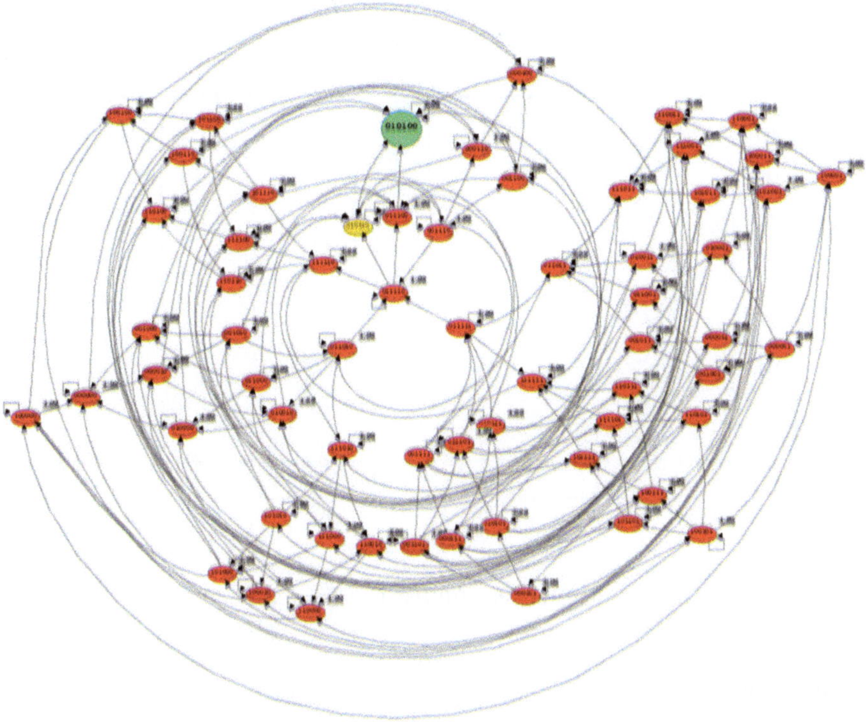

Fig. 3 The state transition graph obtained by applying the GA method

analysis seems to converge with our results. The fixed point 010100 implicates a high concentration of FGF23 and PTH and low concentration of all other components. Theoretical and experimental results show that a low calcium level leads to PTH activation (state 000100), which in turn activates the VD and FGF23 production (state 010101). Increased level of VD also increases the level of Klotho, acting as a co-receptor with FGF23 in renal calcium retention. In the next step, increased level of FGF23 leads to several negative feedback loops (state 111000). The same result is obtained even if we analyze the concentrations of each node, when starting from a hypocalcitrosis state (Fig. 4). In the first time steps, the system oscillates chaotically then finds the fixed point and remains in this state.

The obtained results show that, for small biological networks, even the synchronous updating method can give acceptable results, because of the low variety of the processes involved within the network. Even so, the developed asynchronous updating methods can highlight a numerous characteristics of the network, including absorption time, absorption probabilities and including additional mathematical techniques to study its stability.

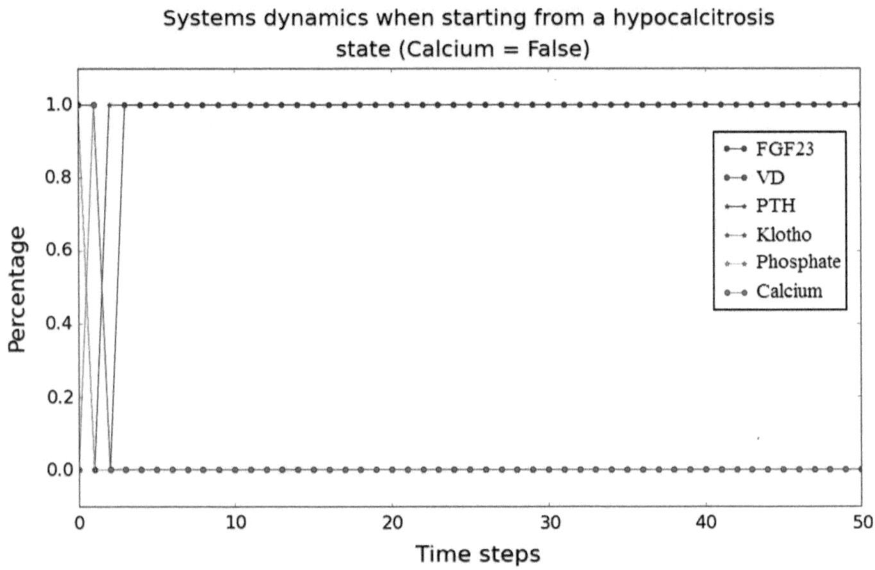

Fig. 4 System' dynamics starting from state 010100, for 50 time steps

References

1. Trairatphisan, P., et al.: Recent development and biomedical applications of probabilistic Boolean networks. Cell Commun. Signal. **11**(1) (2013)
2. Chaves, M., Sontag, E.D., Albert, R.: Methods of robustness analysis for Boolean models of gene control networks. arXiv:0605004 (2006)
3. Saadatpour, A., Albert, I., Albert, R.: Attractor analysis of asynchronous Boolean models of signal transduction networks. J. Theor. Biol. **266**(4), 641–656 (2010)
4. Saadatpour, A., Albert, R.: Boolean modeling of biological regulatory networks: a methodology tutorial. Methods **62**(1), 3–12 (2013)
5. Kauffman, Stuart A.: The Origins of Order: Self Organization and Selection in Evolution. Oxford University Press, USA (1993)
6. Thomas, R.: On the relation between the logical structure of systems and their ability to generate multiple steady states or sustained oscillations. In: Numerical Methods in the Study of Critical Phenomena, pp. 180–193. Springer, Berlin, Heidelberg (1981)
7. Shmulevich, I., et al.: Probabilistic Boolean networks: a rule-based uncertainty model for gene regulatory networks. Bioinformatics **18**(2), 261–274 (2002)
8. Shmulevich, I., Dougherty, E.R., Zhang, W.: From Boolean to probabilistic Boolean networks as models of genetic regulatory networks. Proc. IEEE **90**(11), 1778–1792 (2002)
9. Mestl, T., Plahte, E., Omholt, S.W.: A mathematical framework for describing and analysing gene regulatory networks. J. Theor. Biol. **176**(2), 291–300 (1995)
10. Marku, M., Ifti, M., Koçiaj, I., Klotilda, N.: Dynamical Analysis of Synchronous Boolean Model of the Klotho Gene Complex Networks and Their Applications (2016)
11. Kuro-o, M.: Klotho and aging. Biochimica et Biophysica Acta (BBA)-General Subjects, vol. 1790, no. 10, pp. 1049–1058 (2009)
12. Quarles, L.D.: Role of FGF23 in vitamin D and phosphate metabolism: implications in chronic kidney disease. Exp. Cell Res. **318**(9), 1040–1048 (2012)

13. Haussler, M.R., et al.: The role of vitamin D in the FGF23, klotho, and phosphate bone-kidney endocrine axis. Rev. Endocr. Metab. Dis. **13**(1), 57–69 (2012)
14. Wiese, R., Eiglsperger, M., Kaufmann, M.: yfiles-visualization and automatic layout of graphs. In: Graph Drawing Software, pp. 173–191. Springer, Berlin, Heidelberg (2004)
15. Albert, I., et al.: Boolean network simulations for life scientists. Source Code Biol. Med. **3**(1) (2008)

Author Index

B. Gonçalves et al. (eds.), *Complex Networks VIII*,
Springer Proceedings in Complexity, DOI 10.1007/978-3-319-54241-6

MIX
Papier aus verantwortungsvollen Quellen
Paper from responsible sources
FSC® C105338
FSC
www.fsc.org

If you have any concerns about our products,
you can contact us on
ProductSafety@springernature.com

In case Publisher is established outside the EU,
the EU authorized representative is:
Springer Nature Customer Service Center GmbH
Europaplatz 3, 69115 Heidelberg, Germany

Printed by Libri Plureos GmbH
in Hamburg, Germany